Contents

Introduction

Today we live in a world that is dominated by machines. In the home, machines vacuum the carpet, wash the dishes, and launder the clothes. On the farm, machines plow the land, sow the seeds, and reap the harvest. In industry, machines spin cloth, hammer metal into shape, hoist up heavy loads, and dig minerals out of the ground. In transportation, vehicles of every kind – from bikes and cars to trucks and locomotives – transport us and our goods from place to place.

Exactly what is a machine? Scientists tell us that a machine is a device that performs work. In a more popular sense, a machine is a device that helps us carry out a particular task more easily and usually much faster.

For example, it is difficult for us to till the soil with only our hands. It is much easier if we use a spade. A spade is a simple machine. It is even easier to till the soil if we use a powered digging machine, like a rototiller, which has an engine that drives rotating blades. Usually, when we use the term machine, we mean a mechanical device driven by a motor or an engine.

In this book, we investigate some of the principles behind the machines we use. We also look at engines – the machines that drive other machines, and some typical machines at work.

You can check your answers to the questions featured throughout this book on pages 60-61.

The steam locomotive is among the many magnificent machines that have helped shape our modern way of life.

1

Simple Machines

◀ **The big wheel at a fairground. The wheel Is a simple device that features, in one form or another, In almost every other machine.**

In the Introduction, we gave the spade as an example of a simple machine. It hardly seems to be a machine at all, but it does have something in common with all machines – it helps you perform a task – digging – more easily than you could without it.

More technically, the spade increases the force you put into the digging action. To turn over the soil, you push down on the handle. We call this force the effort. The spade is designed so that it magnifies this effort into a more powerful force, which lifts up a spadeful of soil (a weight called the load).

The spade is an example of a kind of simple machine called the lever. In this chapter, we look at the principles behind the lever and the five other kinds of devices scientists recognize as simple machines. They are the inclined plane, the wedge, the screw, the wheel and axle, and the pulley. We also look at gears, which we can think of as useful modifications of the wheel.

▶ **Gardeners dig the ground with a lever, which we know better as a spade. The handle gives them leverage to move a spadeful of soil.**

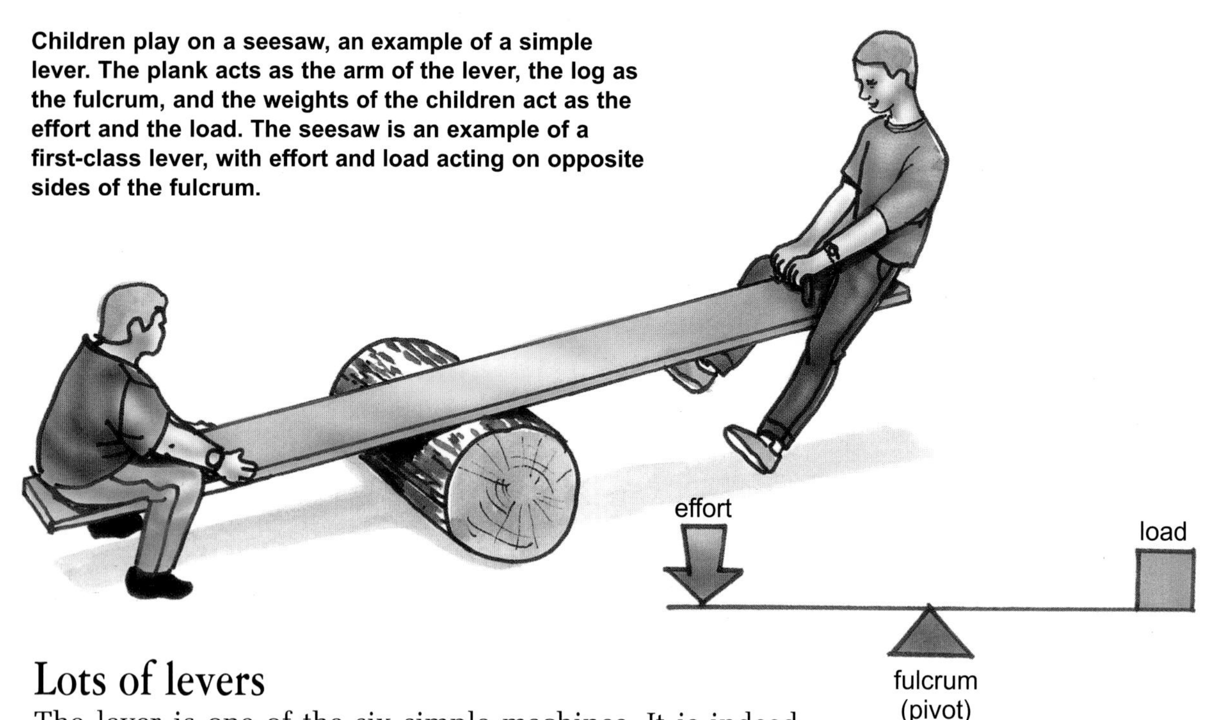

Children play on a seesaw, an example of a simple lever. The plank acts as the arm of the lever, the log as the fulcrum, and the weights of the children act as the effort and the load. The seesaw is an example of a first-class lever, with effort and load acting on opposite sides of the fulcrum.

Lots of levers

The lever is one of the six simple machines. It is indeed simple, being just a straight rigid pole. Yet, it can still perform useful work.

A seesaw is an example of the simplest kind of lever. It consists of a long plank, which rests on a pivot in the middle. If you sit on one end of the seesaw, you exert a downward force, or effort on this end. The seesaw converts this effort into a force acting upward at the other end of the plank. This upward force can lift a friend sitting on that end. His or her weight forms the load on the seesaw.

In general: a lever rests on a pivot, also called a fulcrum. A force called the effort is applied at one point on the lever in order to overcome another force, the load, at another. Usually, a lever in some way magnifies the effort to move a bigger load. The ratio of load to effort is called the mechanical advantage.

Classes of levers

In the seesaw, the effort is applied on one side of the pivot to lift a load on the other side. This arrangement is typical of one kind of lever, called the first-class lever. In other classes of levers, the effort, pivot, and load lie in different relative positions.

In a second-class lever, the pivot is located at one end.

The effort is applied at the other end to lift a load in the middle. The wheelbarrow is a second-class lever. It has the pivot at one end and the load in the middle. A bottle-opener is another second-class lever.

Your forearm is an example of a third kind of lever, the third-class lever. It has the pivot at one end - the elbow. The load is held in the hand at the other end. The effort, which is supplied by the muscle in the arm, acts in between.

When you use a heavy crowbar to lift a heavy packing case, you have to push down the handle a long way to lift up the packing case a little. By pushing down a long way with a small force (effort), you have created a much larger force that can lift the heavy weight (load). We say that the long handle of the crowbar has given you leverage.

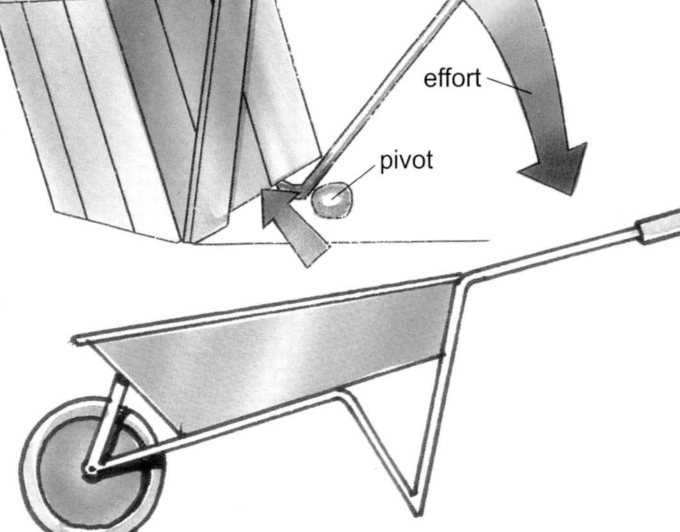

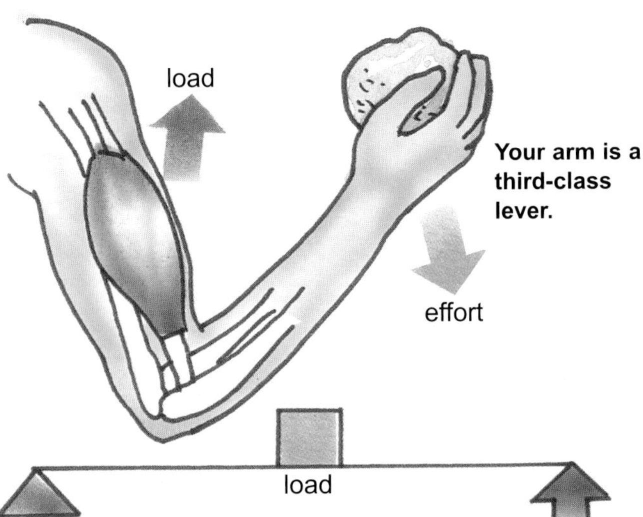

Your arm is a third-class lever.

The wheelbarrow is a second-class lever.

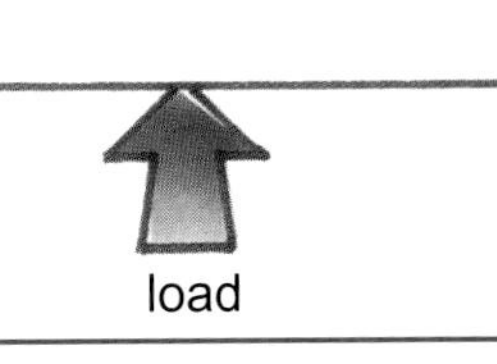
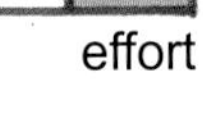

Levering the world

A famous Greek scientist named Archimedes realized that you could lift heavy loads by using a suitable lever. He is supposed to have said: "Give me a place on which to stand, and I will move the world."

Double levers

Pliers, nutcrackers, and sugar tongs are examples of double levers. Pliers are a double first-class lever, with the pivot near one end. Nutcrackers and sugar tongs are double levers with the pivot at one end.

Q 1: Why aren't pliers designed the other way around, with small handles and large jaws?

Q 2: Make sketches of a pair of nutcrackers and a pair of sugar tongs, showing where the effort and load act when you use them. Looking at your sketches, fill in the missing words in the following: Nutcrackers are a double ...... class lever; sugar tongs are a double ..... class lever.

Slopes and wedges

It might be difficult to imagine that there would be a simpler machine than the lever, but there is. It is a slope or, as scientists call it, an inclined plane. You can use a plank as an inclined plane to help move loads more easily. Suppose you are pushing a heavy wheelbarrow and you come to a steep step.

INVESTIGATE

Lever rules

For this investigation, you need a stiff ruler, a pencil, and some coins, such as quarters. Balance the ruler on the pencil. You now have a first-class lever. The pencil is the fulcrum. It should be located exactly under the center of the ruler.

Next, place one coin near the edge on one side of the ruler. Place another coin on the other side until the ruler balances (1.). Measure the distances from the pencil to the center of each coin. Are these distances the same, or different?

Next, place another coin on top of the coin on one side of the ruler. This makes the ruler tip up (2.). You can make the ruler balance again by moving the pencil (3.). Do this and then measure the distances from pencil to the coins on each side. What do you find?

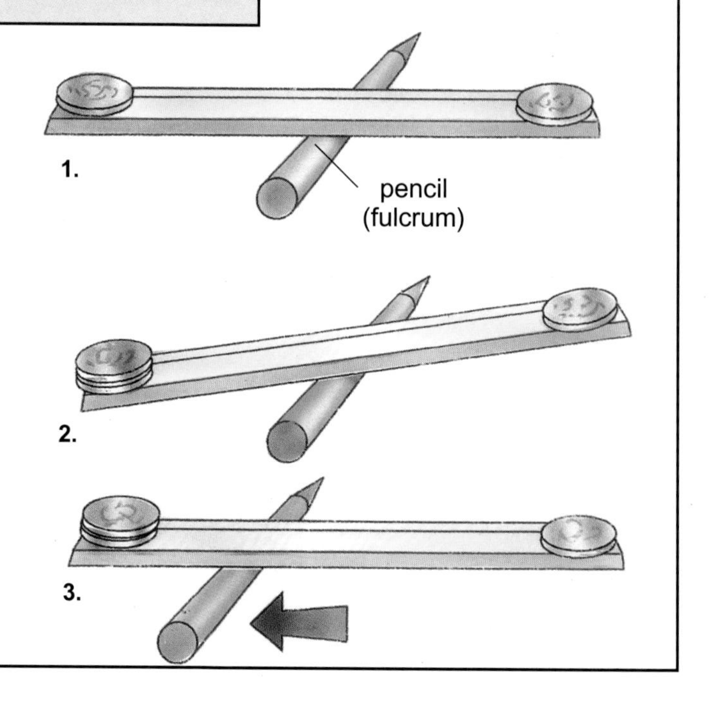

You can't lift the barrow upward because it is too heavy, but if you place one end of a plank on the step and the other end on the lower ground, you can push the barrow up the plank quite easily. In other words, you are using the plank as a simple machine – an inclined plane – to help you do work more easily.

If we put two inclined planes together, back to back, we get another kind of simple machine – the wedge. A wedge used for splitting wood exerts a sideways force as it is driven in. A small force moving the wedge forward a relatively long distance, creates a much greater force moving sideways a relatively short distance.

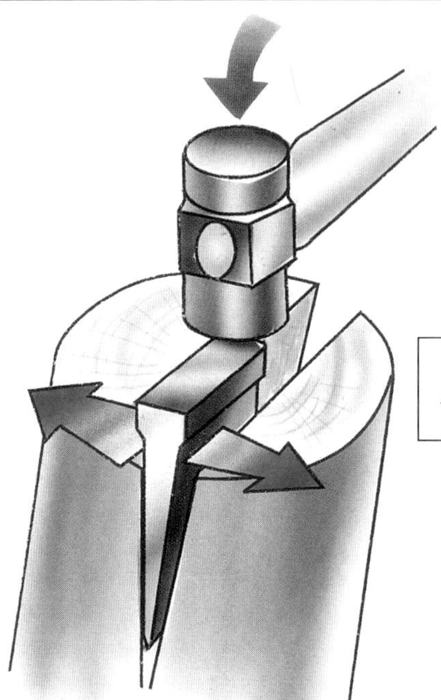

▲ A wedge can be used to split logs. When hit by a hammer, it moves downward into the wood, and it creates powerful forces acting sideways. These forces eventually cause the wood to split.

◀ A skier can travel straight down a mountain slope, but he can't climb straight back up it. He has to zig-zag from side to side, climbing only a little each zig and zag. He is using the principle of the Inclined plane to travel vertically upward with less effort.

Screws

The screw is another kind of simple machine that is very useful. It consists of a body with a spiral thread cut around the outside. Like other simple machines, the screw is designed to produce a bigger force than the force applied to it. This extra force can then push the screw into wood or metal, for example. Screws are one of the main means we use to join things together.

There are many kinds of screws with different kinds of threads. Some threads are coarse – they are wide apart. Some threads are fine – they are close together. The distance between two threads is called the pitch.

Q Find some screws and count the number of threads they have. Does the number vary from screw to screw?

The great advantage

Carry out the Investigation to find the mechanical advantage of a typical screw, that is, the ratio of load to effort (see page 10). You should find it remarkably large. It is for this reason that practical machines using screws can be very useful. One such machine is the screw jack drivers use to lift a car when they want to change a tire. Using this device, they can lift the car, which can weigh more than half a ton, with ease.

Nuts and bolts

Ordinary screws have a tapered body, like a cone. Other screw devices have bodies like cylinders. They are called bolts. They are used widely to join parts of machines together. They may screw directly into metal, or into nuts. Nuts are metal pieces with screw threads on the inside. When a bolt screws into a nut, powerful forces are produced that squeeze together the parts to be joined.

Below: A nut and bolt exert powerful forces when they are tightened. Most nuts and the head of most bolts are hexagonal In shape. They are tightened using a wrench.

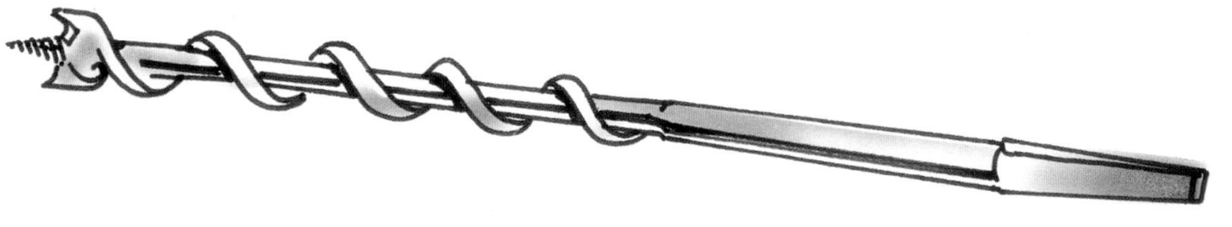

Bottom: A carpenter's drill bit has a fine screw thread at the tip and a very coarse thread on the body.

▶ By turning the handle of the screw jack many times, you are able to lift a heavy load, such as a car. You are applying a small effort a long distance to lift a heavy load a short distance.

▼ This device was one of the first practical machines. The Greek Archimedes invented it in the 200s BC, and it is called the Archimedean screw.

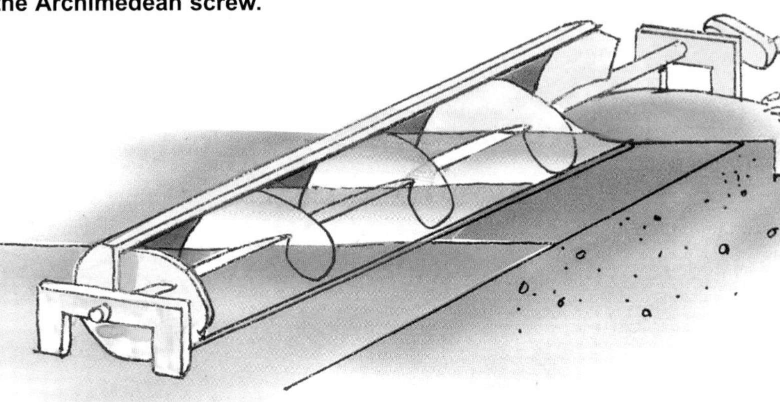

INVESTIGATE

Finding the advantage

Like most machines, a screw magnifies the force (effort) you put into it, and the magnified force can lift a load. The ratio of load to effort is the mechanical advantage.

We can work out from the dimensions of a screw how much it increases the force you put into it. When you turn the screw once, it moves forward into a piece of wood, say, the distance between two threads. The amount the screw increases a force is given by the circumference of the screw, divided by the width of the threads.

Using a large wood screw, measure the circumference of the head by rolling it on a sheet of paper, as in (1). Now measure the distance between the threads (2). It is best to measure the distance of a number of threads and then divide by the number.

When you have taken your measurements, divide the first by the second, and this will give you the force-multiplying effect, or mechanical advantage of the screw. Repeat the investigation using other screws.

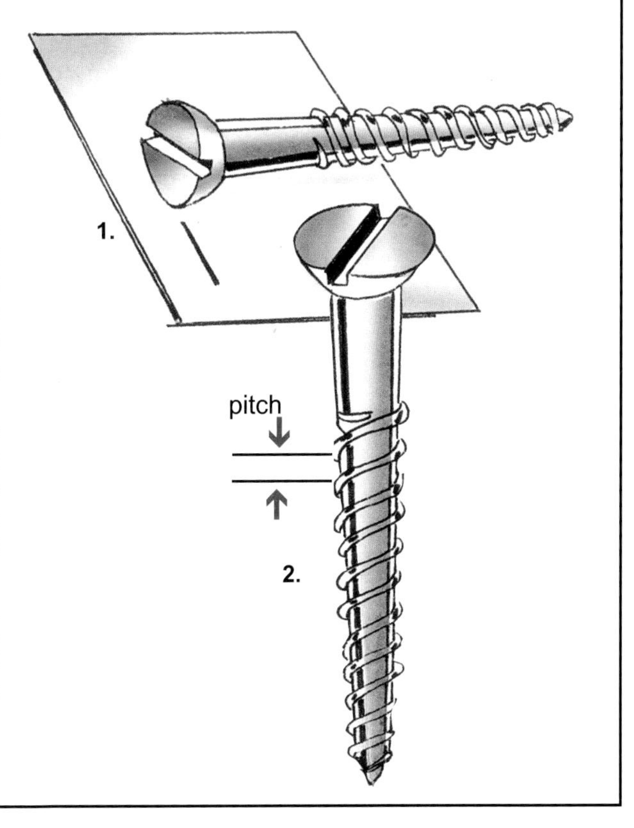

Powerful wheels in the form of a modern dragster. From a standing start, dragsters can cover a quarter-mlle in less than 5 seconds. By this time it could be traveling at a speed of 300 mph (480 km/h) or more.

Wheels and axles

Ask your friends what they think is humankind's greatest invention, and make a note of what they choose. They might say, for example, the laser, the space rocket, the silicon chip, the computer, or the television. But arguably it is none of these. It is something much simpler – the wheel.

The invention of the wheel is thought to have taken place about 5,500 years ago. Today, wheels in one form or another are found in most vehicles and engines and in other machines and gadgets of all kinds, from wristwatches to locomotives.

Q How many machines or devices can you think of that use wheels in one form or another?

Wheel machines

Wheels form part of hundreds of different machines, but they are not machines in themselves. A wheel becomes a machine when it is mounted on an axle. It becomes a simple machine called a wheel and axle. It is a machine because it can do work.

A windlass is a kind of wheel and axle used to lift loads. For centuries it has been used to raise buckets from water wells. In the windlass (right), a rope carrying the bucket of water passes over an axle. The axle is turned by a wheel with a handle.

When you turn the handle, your hand moves in a large circle, while the axle turns in a small circle. This action turns the small force you apply to the handle into a bigger force at the axle. This bigger force winds up the rope with its heavy load.

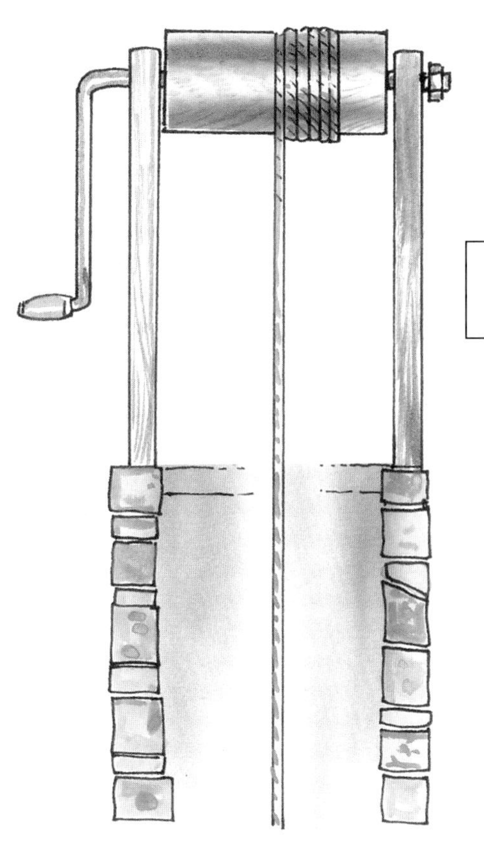

A windlass Is an age-old device used to lift buckets of water from a well.

INVESTIGATE

Rolling along

Why use wheels? Stack three bricks on one another. Tie a string around the bottom one, and pull on it. Is it easy or difficult to move the stack of bricks in this way?
Now place the bottom brick on a row of round rods or pencils, and stack the other two bricks on top. Try pulling the stack of bricks now. Is it easier or more difficult to pull than before?

More wheels and axles

How many wheels are there on a bicycle? Two. wrong! There are three. Apart from the two big wheels, there is the small toothed wheel to which the pedals are attached. It is called the chain wheel because it moves the chain when you pedal.

The chain wheel and pedal system is another example of that simple machine, the wheel and axle. When you pedal your bike, the pedals (acting as the wheel) turn in a large circle, while the chain wheel (acting as the axle) turns in a small circle. This means that the force you apply to the pedals is turned into a bigger force at the chain wheel.

Cyclists use the mechanical advantage of the wheel and axle when they pedal their bikes.

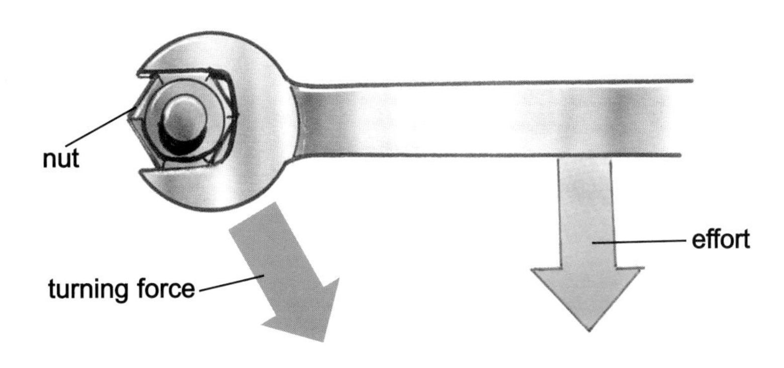

When you tighten a nut with a spanner, you use the mechanical advantage of the wheel and axle to magnify your effort.

Pulleys

Another simple machine using a wheel is the pulley. It is used to lift loads. A single pulley consists of a wheel on an axle with a groove around the rim. A rope passes over the wheel, resting in the groove. In operation, the pulley axle is fixed to a beam, and a rope tied to the load passes up and over the pulley. You pull on the free end of the rope to lift the load. Using this single pulley doesn't give any increased lifting force, or mechanical advantage.

But if you use two or more pulley wheels with your rope, you do get a mechanical advantage. A system with more than one pulley is called a pulley block. Together with the rope (tackle), it is called a block and tackle. Pulley blocks are widely used in industry for lifting heavy loads. Most cranes have a pulley block attached to the lifting hook

pulley wheels
axles
pulley wheel
effort
load

This is an example of a pulley block with three pulley wheels. It has a mechanical advantage of three – magnifies your effort three times. If you added another two pulleys, your effort would be magnified six times.

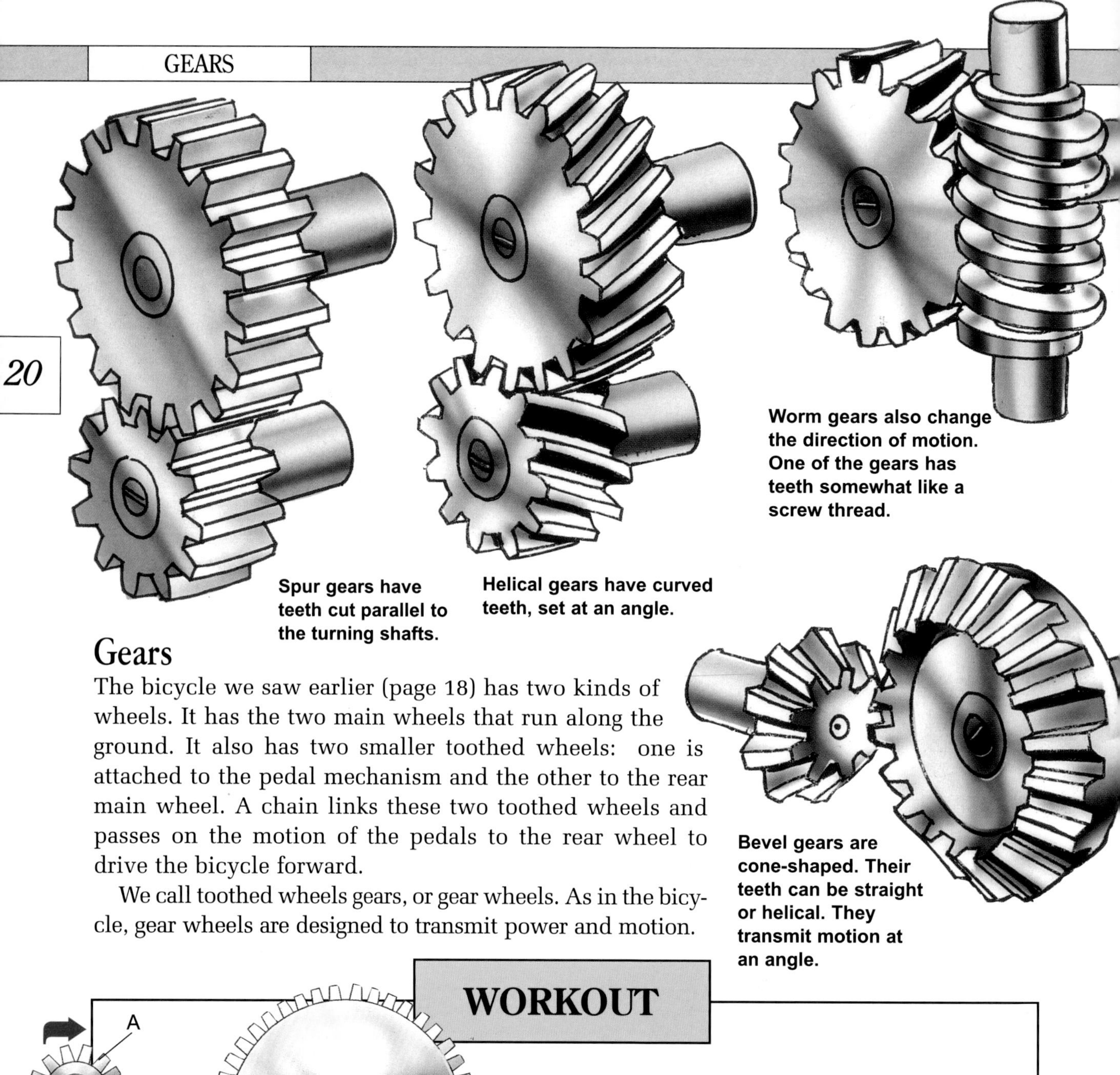

Gears

The bicycle we saw earlier (page 18) has two kinds of wheels. It has the two main wheels that run along the ground. It also has two smaller toothed wheels: one is attached to the pedal mechanism and the other to the rear main wheel. A chain links these two toothed wheels and passes on the motion of the pedals to the rear wheel to drive the bicycle forward.

We call toothed wheels gears, or gear wheels. As in the bicycle, gear wheels are designed to transmit power and motion.

WORKOUT

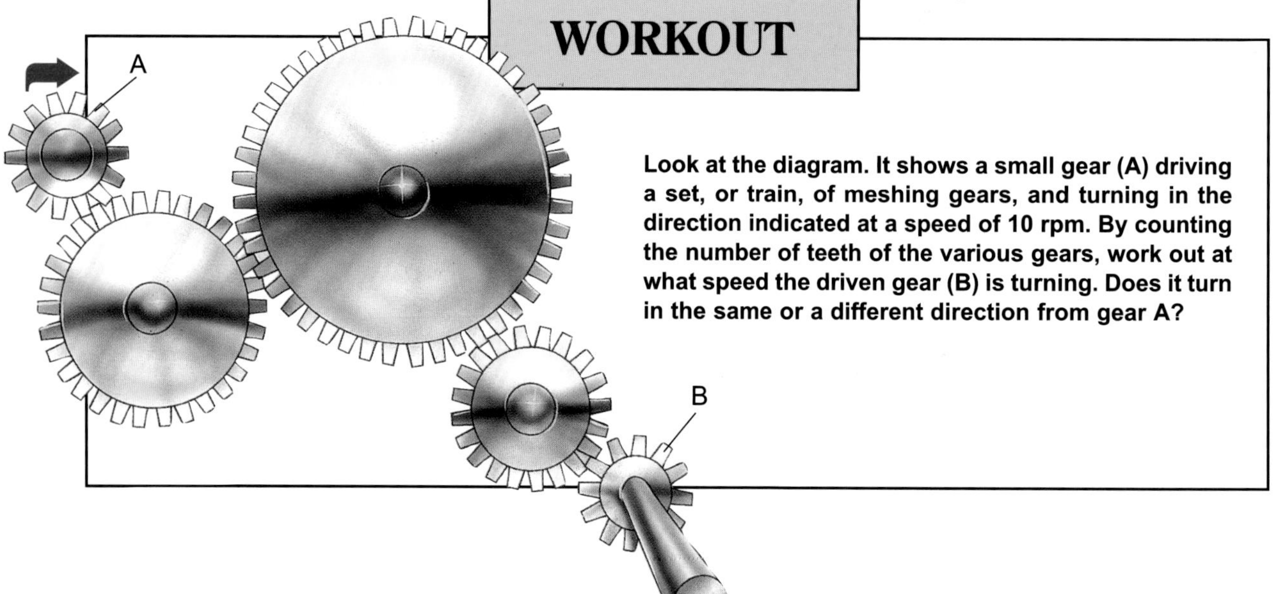

Look at the diagram. It shows a small gear (A) driving a set, or train, of meshing gears, and turning in the direction indicated at a speed of 10 rpm. By counting the number of teeth of the various gears, work out at what speed the driven gear (B) is turning. Does it turn in the same or a different direction from gear A?

Kinds of gears

There are several kinds of gears, which can transmit power to shafts that are parallel or at different angles to each other. The simplest kind of gear is the spur gear, which has straight teeth. A helical gear has curved teeth, and it is smoother and quieter in operation than a spur gear. Bevel gears have teeth that slant at an angle, and they transmit motion between shafts at an angle with each other.

Changing speed

If two meshing gears have the same number of teeth, they turn at the same speed. If the driving gear has more teeth than the driven gear, it will drive the driven gear faster. On the other hand, if the driven gear has more teeth, then it will be driven around more slowly.

Gears that increase speed are called multiplying gears. Those that reduce speed are called reducing gears. Reducing gears are used in the transmission system of cars, for example (see page 47). When a car is traveling at about 55 mph (88 km/h), the wheels make about 750 revolutions per minute (rpm). A car engine turns at about 3,000 rpm.

Q What is the missing number in the following statement? The gears must reduce the speeds transmitted by a ratio of _ : 1.

Variable gears

Car engines always need to turn at a high speed, whether the car is crawling or traveling fast. They therefore have a system of variable gears so that the engine can drive the car at different speeds. These variable gears are housed in a gearbox. As the driver changes gear, different sets of gears are brought into mesh to change the high engine speed into a speed suitable for driving the car on the road.

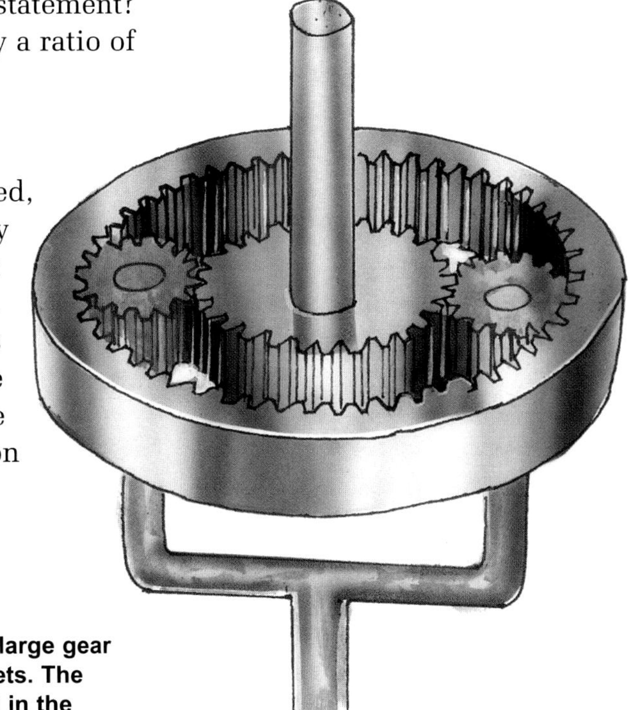

▶ This arrangement is called a sun and planet gear. The large gear at the center is the sun, and the small gears are the planets. The toothed ring is the annulus. This kind of gearing is found in the gearboxes of cars with automatic transmissions.

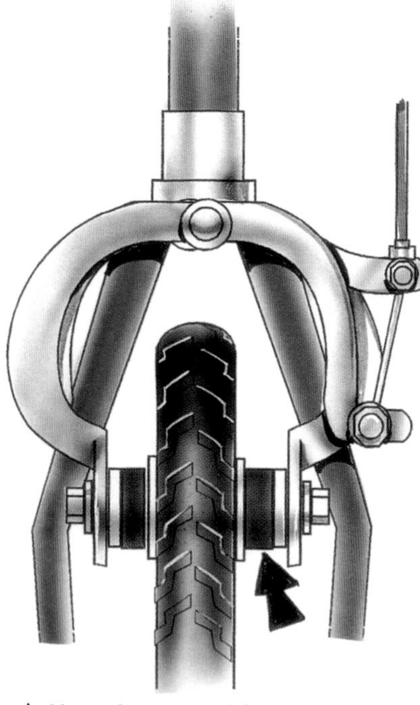

▲ **Next time you ride a bike, pedal hard and then put on your brakes. Now feel the brake blocks. What do you find?**

Q 2. How do you explain what you find?

Bearings

Every practical machine has moving parts. Gears turn on shafts and wheels turn on axles, for example. Moving parts present one major problem. When they move, they usually rub against parts that aren't moving. A wheel turning on an axle rubs against the axle. This rubbing sets up a force that tries to prevent the wheel from turning. It is the force of friction.

In a machine with moving parts, much of the power put into the machine is used up in overcoming friction. The energy absorbed by friction reappears as heat.

Q 1. What would happen in a machine with moving parts if the heat caused by friction wasn't removed? Remember that most machines are made of metal.

Oiling

One way we can reduce friction is by oiling, or lubricating, the parts that are rubbing together. The oil coats each part in a thin film. The two films help keep the moving parts apart, preventing them from rubbing together.

INVESTIGATE

Marble bearings

Find two cans that fit together snugly, one on top of the other. Empty the top can, then fill it with sand. Place it on the bottom can and try to spin it around. What happens? Now place some marbles between the top and bottom cans and try to spin it around again. What happens?

Q 3. What does this investigation tell you about sliding friction and rolling friction?

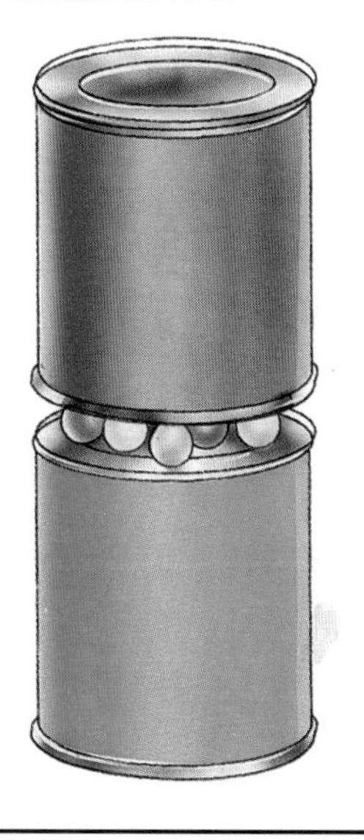

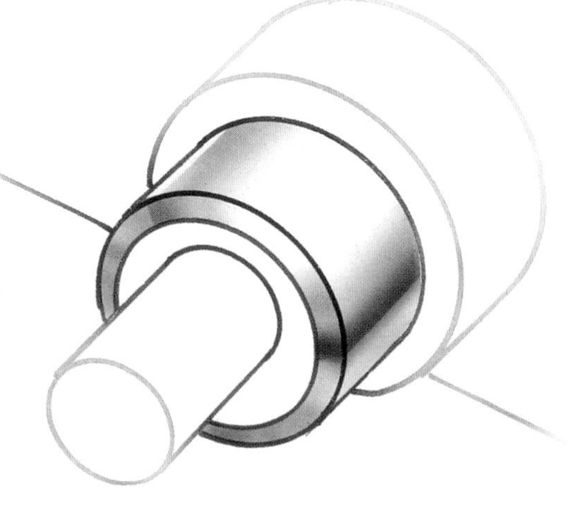

▲ **The main bearings in a car engine are journal, or sleeve, bearings.**

Bearing up

In many machines, the friction between moving parts is further reduced by using bearings. The simplest bearing is the journal bearing, which is a smooth sleeve supplied with oil.

Most bearings are made up of rings of steel balls or rollers. The balls and rollers are designed to roll when the moving parts turn. When an object rolls, it experiences much less friction than when it slides. This is why the wheel works.

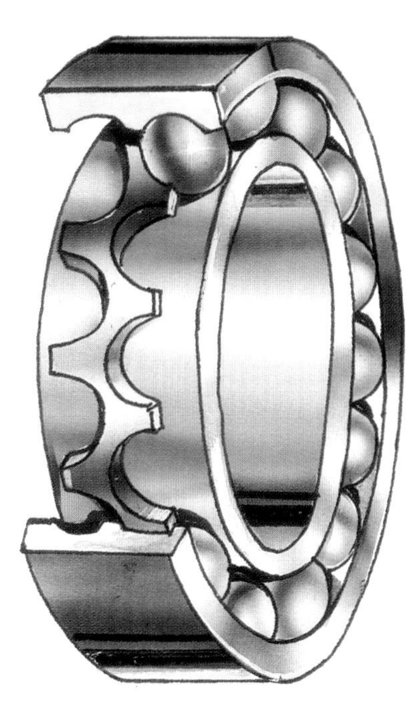

▲ **Ball bearings support the shaft in an electric drill.**

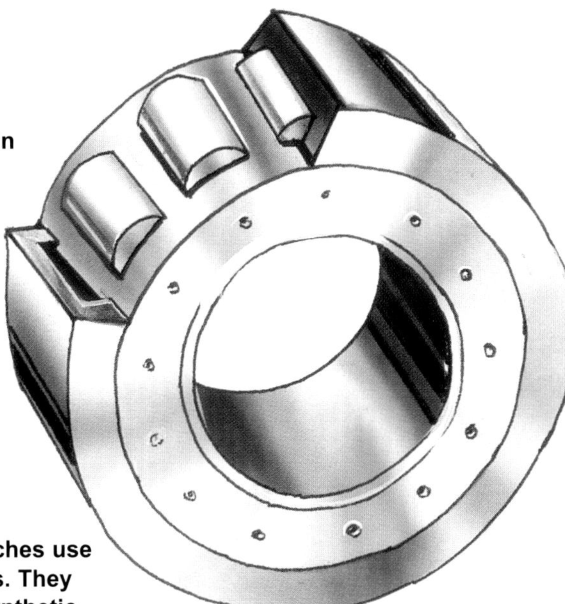

▶ **Roller bearings support the axles in many vehicles.**

▼ **Mechanical watches use jewels for bearings. They are made out of synthetic ruby.**

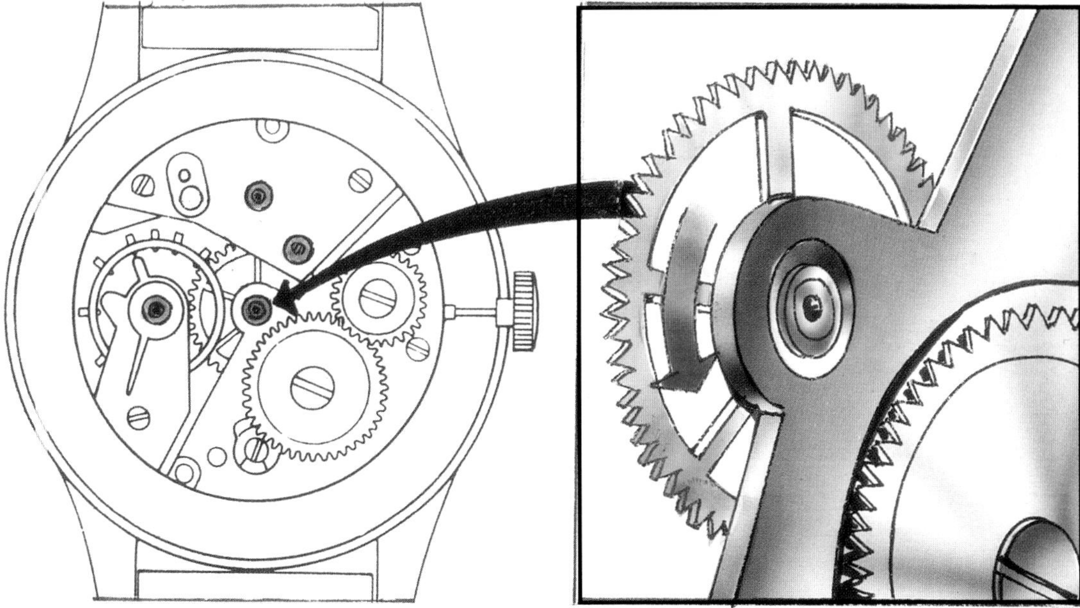

ROLLS-
ROYCE
ROLLS-
ROYCE

2
Engines

◀ **Close-up of a veteran Rolls Royce car engine. Visible near the top are two of the spark plugs that explode the gasoline fuel the engine runs on. The engine harnesses the energy released by the explosion and converts this into mechanical motion to drive the car.**

Practical machines are much more complicated than the simple machines discussed in the first chapter. If you look at them carefully, you will find that they are combinations of parts that are simple machines, particularly levers, wheels, and gears. All these parts work together to perform a particular task, such as making a car run.

We can think of a car as being two machines in one. The first machine is the engine. This changes the energy in a fuel (usually gasoline) into motion. The second car machine is the transmission, which transmits the motion produced by the engine to the wheels that drive the car.

Machines, like a gasoline engine, that produce motion are called prime movers. We use several other kinds of prime movers, such as steam engines, jet engines, and turbines. Electric motors are also very widely used to drive machines. They work by electricity produced by prime movers in power plants.

▶ **A jet engine propels this plane. It Is mechanically simpler than a car engine, consisting basically of a series of rotating turbine wheels.**

The steam engine

Before gasoline engines came into use, the most important engine was the steam engine. Steam engines are piston engines. They consist of a piston that can move back and forth in a cylinder. Steam is introduced one side of the piston and forces it along the cylinder. It may then be introduced to the other side of the piston, forcing it back. Connecting rods pass on the motion of the piston to the machinery being driven. Often the piston motion drives around a heavy wheel. This wheel is called a flywheel. It serves to store the energy produced.

Steam engines are now only of historic interest, but they are extremely important because they provided the power that started the growth of modern industry. Today, different kinds of steam machines are used to produce power. They are steam turbines (see page 31).

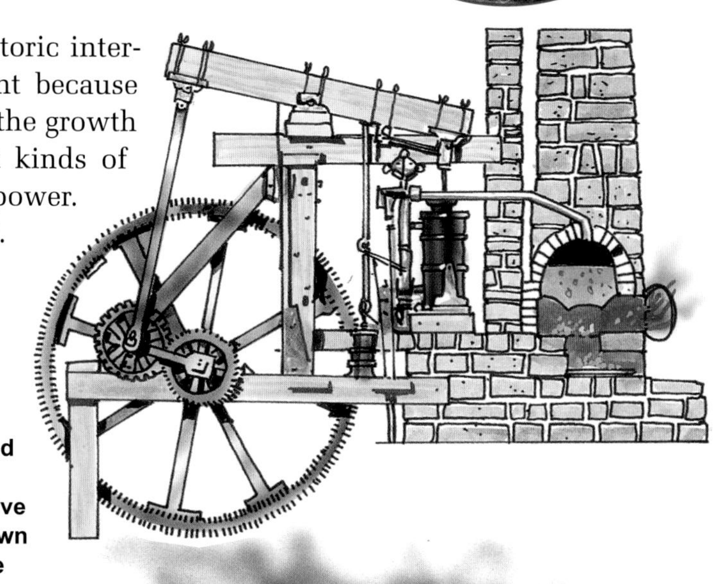

Top right: In 1769, the Scottish engineer James Watt turned the steam engine into a source of power for industry.

▶ One of Watt's beam engines. The engine piston was connected to one end of a pivoted beam. The other end of the beam was connected to a wheel, which was used to drive machinery. When steam drove the piston down the cylinder, the beam rocked and turned the wheel.

Steam engines powered most locomotives until the middle of the 20th century. The pistons moving back and forth in cylinders drove the wheels through a system of rods and cranks.

The gasoline engine

The gasoline engine that powers most cars is also a piston engine. It has pistons moving up and down inside cylinders. Gasoline is a fuel that is burned inside the engine cylinders to produce hot gases. The hot gases force the pistons down the cylinders. The pistons drive around a shaft called the crankshaft, producing the turning power that other parts of the car carry to the driving wheels.

Because fuel is burned inside the engine, the gasoline engine is called an internal combustion engine. Gasoline is obtained by processing crude oil.

Q What is another term for crude oil. What does this word literally mean?

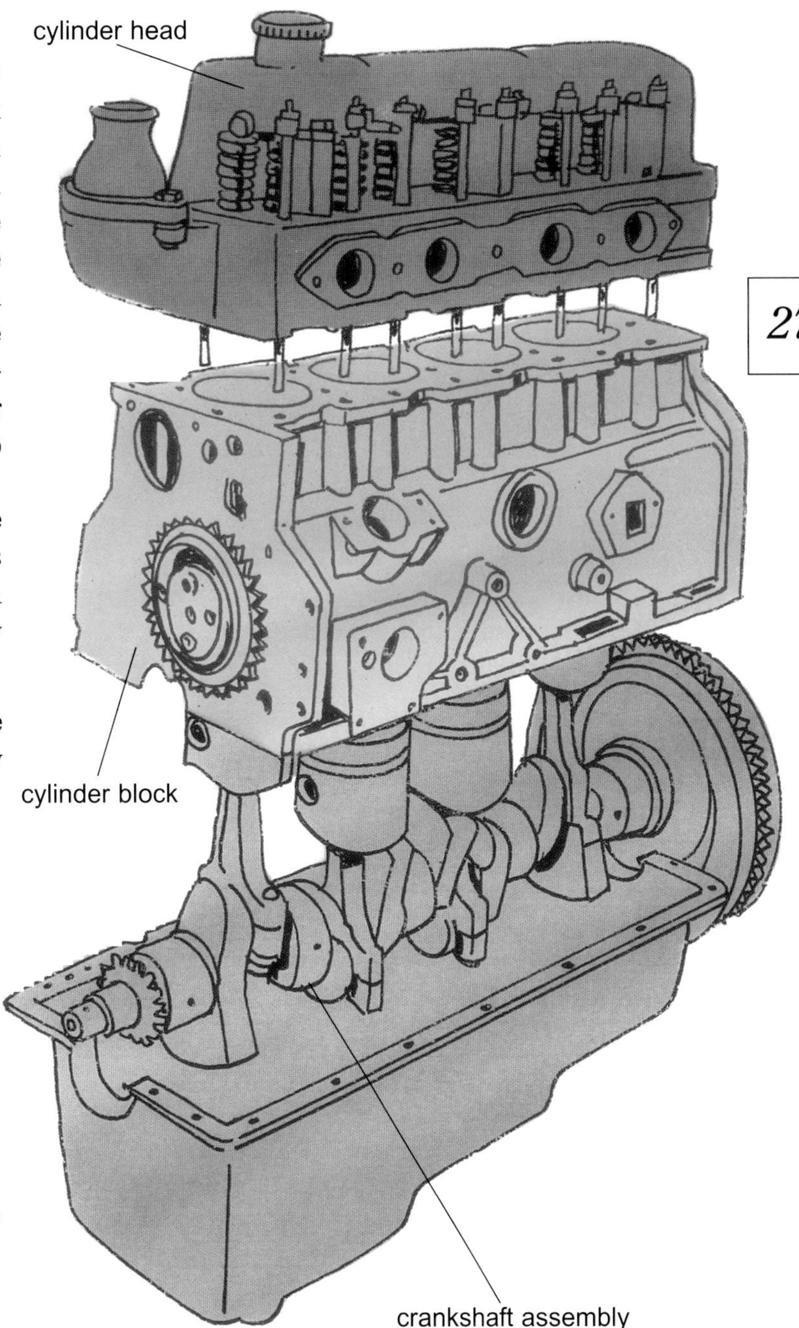

This diagram shows an exploded view of a gasoline engine. It is made up of three main units.

Cylinder head
This unit carries valves that open to allow fuel mixture into the engine cylinders and burned gases out.

Cylinder block
This is the largest and heaviest part of the engine and holds the cylinders. Water circulates through passages in the block to carry away the heat produced when the fuel burns.

Crankshaft assembly
This is the heart of the engine, consisting of the pistons, connecting rods, and crankshaft. At one end of the shaft is a heavy flywheel, which helps smooth out the engine motion.

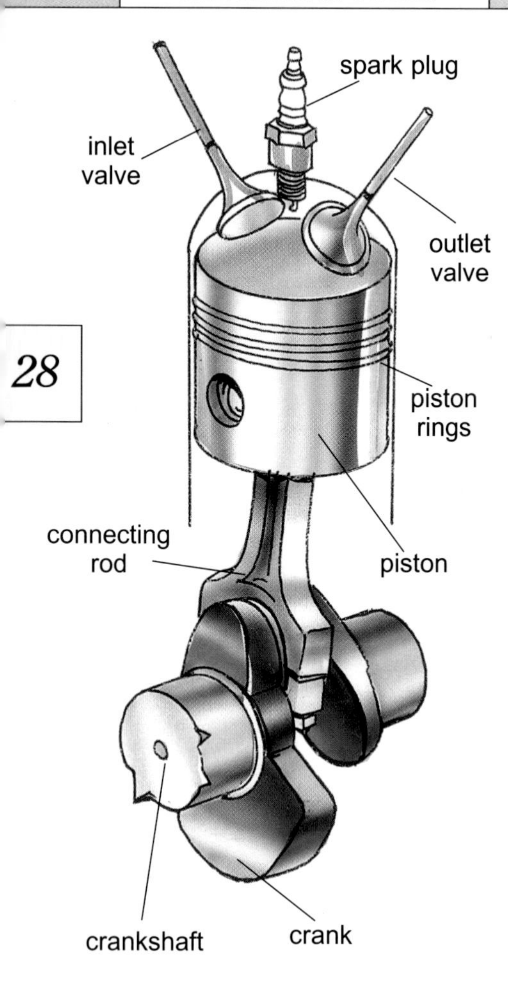

The engine cycle

The diagram shows in more detail the piston/cylinder/crankshaft set-up. The inlet valve opens to allow fuel mixture (gasoline and air) into the cylinder. A spark from the spark plug sets the mixture burning. The outlet valve opens to let the burned gases escape. The piston rings form a tight seal between the piston and the cylinder walls. The connecting rod joins the piston to a crank on the crankshaft.

The four-stroke cycle

In the cylinder, certain actions are repeated in a regular cycle so that the piston can deliver power. These actions are repeated every four piston strokes (movements), so the cycle is called the four stroke cycle. The stages in the cycle are shown below.

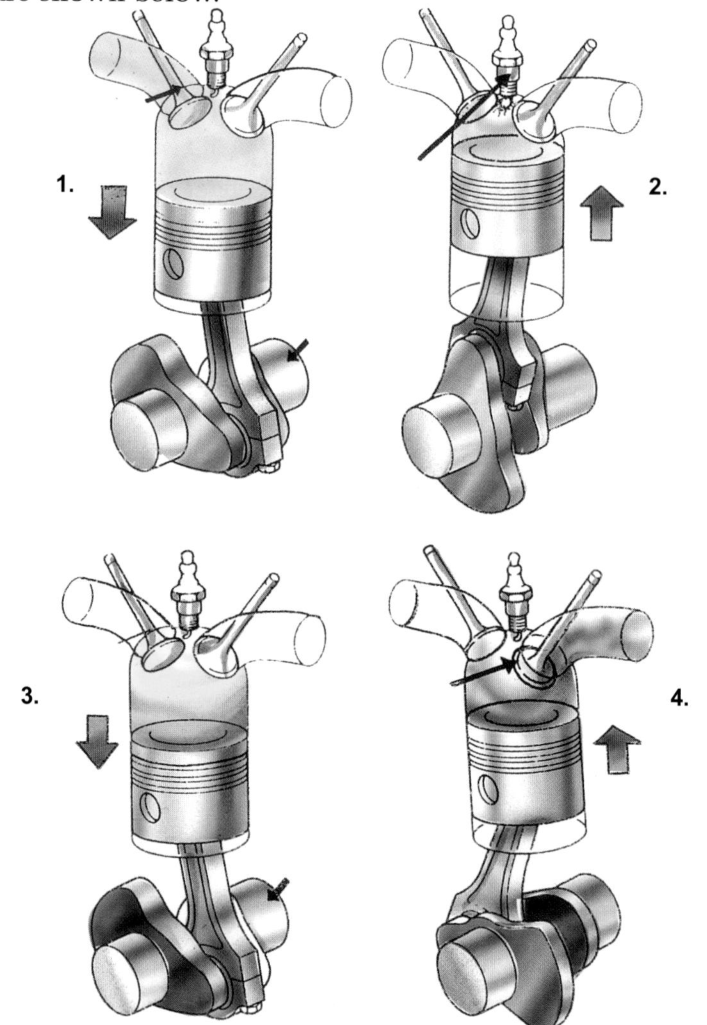

1. Intake
Piston moves down. Fuel mixture is drawn in through open inlet valve.

2. Compression
Inlet valve closes.
Piston moves up, compressing mixture.

3. Power
Spark from spark plug ignites mixture. Hot gases force piston down on its power stroke.

4. Exhaust
Piston moves up, forcing burned gases through open outlet valve.

Operating the valves

For the engine to work properly, the valves must be opened at exactly the right time in the engine cycle. They are opened by means of one or more camshafts.

The action of the camshaft is illustrated at the right. It is a shaft carrying raised parts called cams. It is driven by a chain from the crankshaft. As the camshaft turns, the cams press against the valves and open them. As it continues turning, the raised parts of the cams move on, and springs snap the valves closed.

Engine systems

We have concentrated here on the main mechanisms that make a gasoline engine work. Many other systems are also involved in engine operation. For example, there is the fuel system, which delivers a suitable fuel mixture into the engine cylinders.

There is the ignition system, which provides the spark to the spark plug to ignite the fuel. The cooling system circulates water through the engine to help keep it cool. The lubrication system keeps the moving parts supplied with oil. The exhaust system removes the burned gases from the engine to the rear of the car. It also silences them.

Working on the ignition system of a car. A device called a coil takes low-voltage electricity from a battery and converts It into high-voltage electricity to make a spark.

Turbines

Two of the earliest sources of power for driving machinery were the waterwheel and the windmill. The waterwheel has little paddles or buckets mounted around the edge. When it is placed in a stream, the flowing water spins it around. A windmill has large sails, which turn when the wind blows through them.

Waterwheels and windmills are no longer used for driving machinery today. But we do find uses for modern versions of them – water and wind turbines. Water turbines are the most useful. They are used to harness the energy in flowing water to drive electricity generators in hydroelectric power schemes.

Q What does the word "hydroelectric" literally mean?

▼ **Windmills were a major power source until early this century. They are a simple turbine, harnessing the energy of the wind with their huge sails.**

Propeller turbines

Modern water turbines look nothing like the old waterwheels. They look more like a ship's propeller. Likewise, wind turbines look nothing like the old windmills. Some look like airplane propellers; others look like egg-beaters.

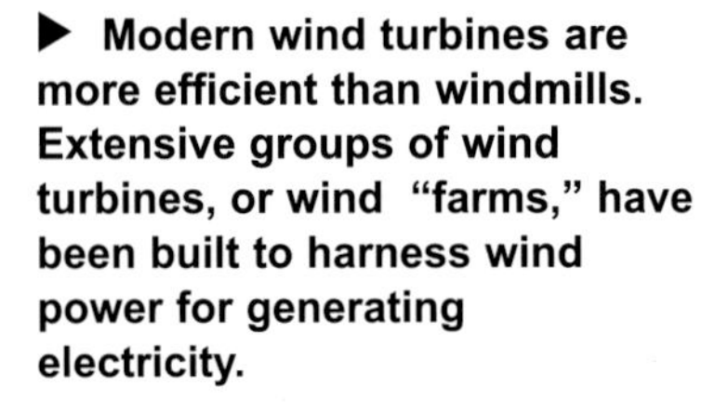

▶ **Modern wind turbines are more efficient than windmills. Extensive groups of wind turbines, or wind "farms," have been built to harness wind power for generating electricity.**

Steam turbines

More important and much more powerful than water and wind turbines are turbines spun by steam. Steam turbines are used in most power plants to drive the generators that produce our electricity. Many large ships use steam turbines to turn their propellers.

A steam turbine consists basically of a shaft (rotor) that can rotate inside a fixed casing (stator). Mounted on the shaft are sets of turbine wheels, which take the form of many-bladed disks. When steam passes through the machine, it makes the rotor spin around.

Steam enters the turbine at a temperature of hundreds of degrees and at a pressure several hundred times the pressure of the atmosphere. After passing through the turbine, the steam is condensed, or turned back into water. This lowers the pressure, which helps "pull" the steam through the turbine faster.

▲ Waterwheels have been in use for about 2,000 years. Old types like this are no longer used. But their descendants, water turbines, are widely used In hydroelectric power schemes.

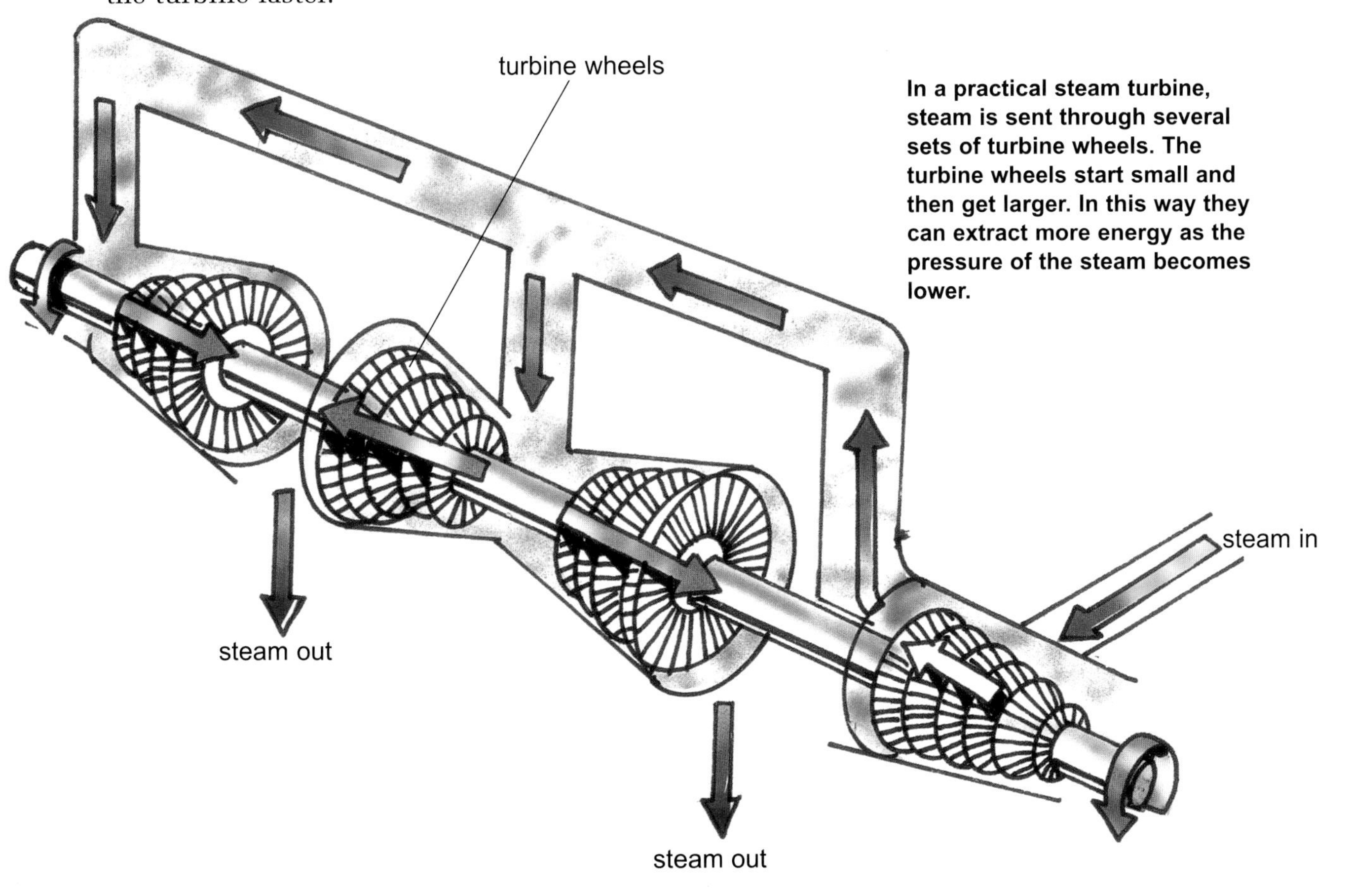

In a practical steam turbine, steam is sent through several sets of turbine wheels. The turbine wheels start small and then get larger. In this way they can extract more energy as the pressure of the steam becomes lower.

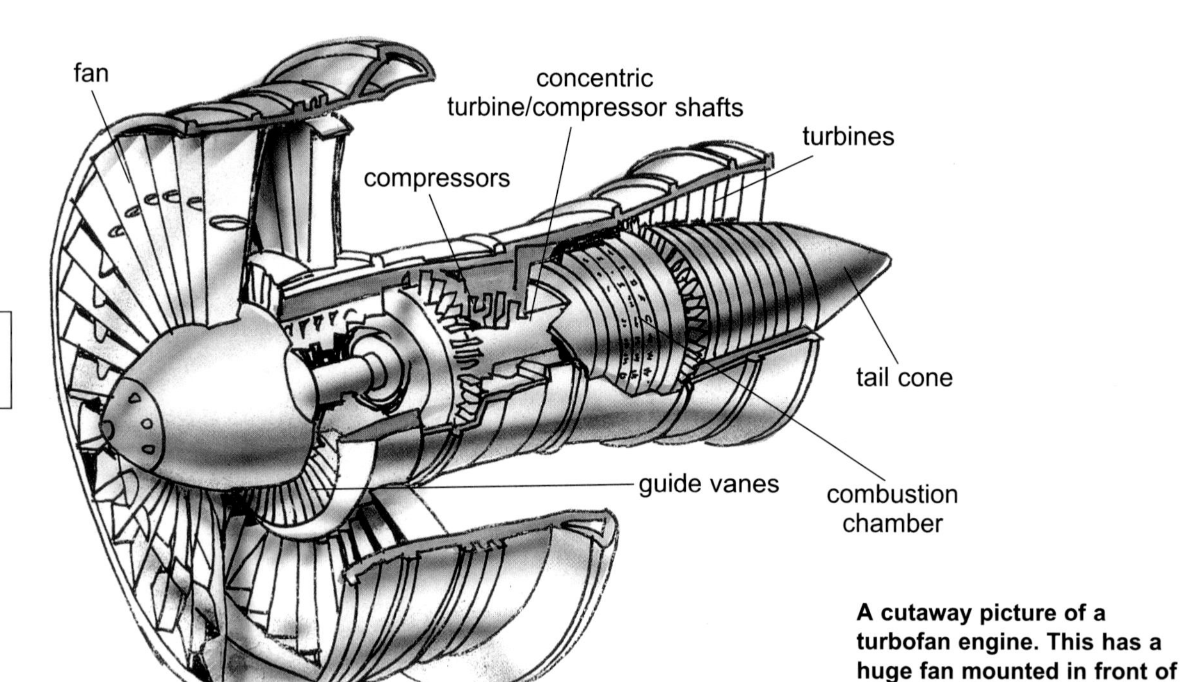

A cutaway picture of a turbofan engine. This has a huge fan mounted in front of the compressor. It forces air not only into the engine but also around It. This "by-pass" air gives the engine added thrust.

Gas turbines

More powerful still than the steam turbine is the gas turbine. In this kind of turbine, fuel is burned to produce hot gases, which spin the wheels on the rotor. Gas turbines are used to spin the generators in some power plants, and some are used elsewhere in industry. But gas turbines are used mostly to power airplanes. They are used in a form we know as the jet engine.

The turbojet

The simplest kind of jet is the turbojet, which is illustrated in the diagram. It consists of a rotor mounted inside a casing. At the front, the rotor carries a set of blades. This part is called the compressor, because when it spins, it sucks in air and compresses it. The compressed air enters the combustion chamber. Fuel is sprayed in and burned. The hot gases shoot out backward and spin the wheels of a turbine. This turbine turns the compressor.

The hot gases continue

The main parts of a turbojet, the simplest member of the jet-engine family. The gases produced by burning the fuel spin a turbine that turns the compressor. They they escape backward to provide thrust.

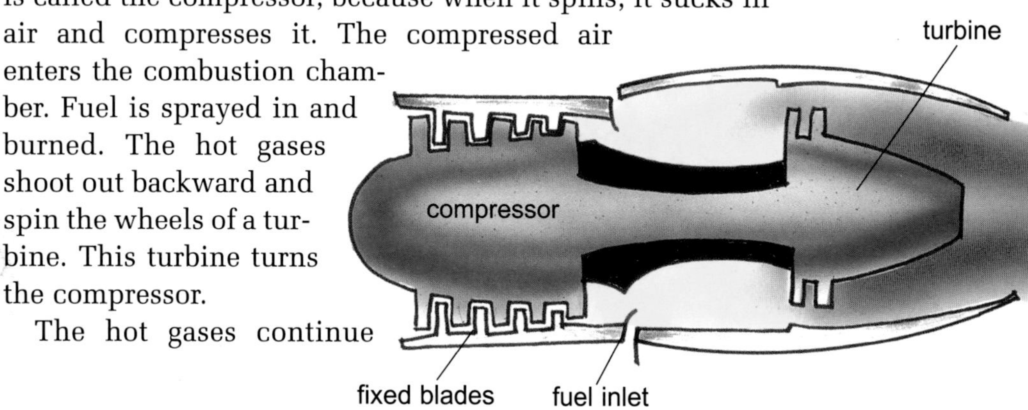

shooting backward and leave the engine through a nozzle as a stream, or jet. As the jet shoots backward, the engine is thrust forward. This effect is known as reaction.

Practical jet engines are more complicated. They have extra compressor and turbine stages. The engine shown at left is a jet called a turbofan. It is so called because it has a huge fan mounted in front of the main compressor. It is more efficient than a turbojet and is the kind used to power most airliners.

Rocket power

The most complicated flying machine of all – the space shuttle – is powered by rocket engines. Rocket engines are very simple in design. They consist of little more than a combustion chamber and a nozzle. Fuel and oxidizer (a substance that provides oxygen, are fed to the chamber and burned. The hot gases produced shoot out backward and drive the rocket forward.

Q Why can't the space shuttle use jet engines to fly into space?

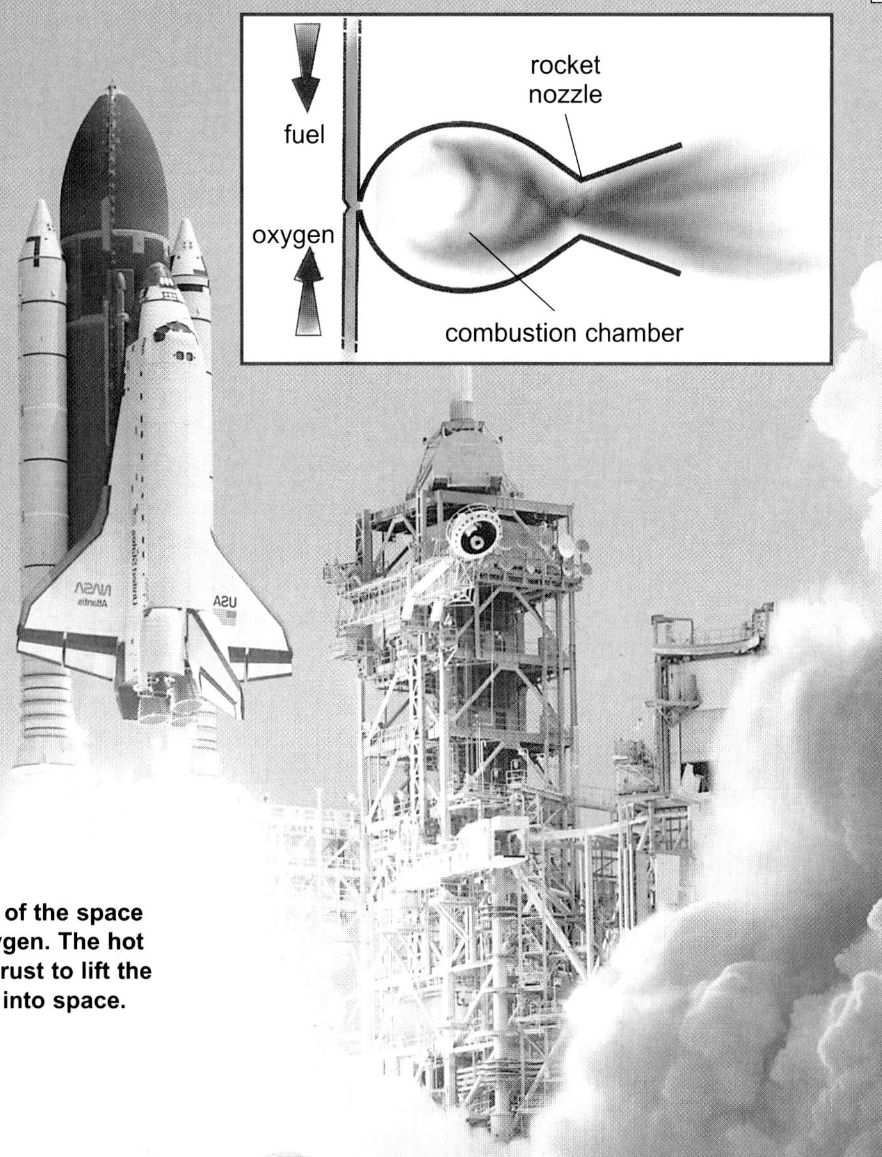

The three main rocket engines of the space shuttle burn hydrogen and oxygen. The hot gases produced provide the thrust to lift the vehicle off the launch pad and into space.

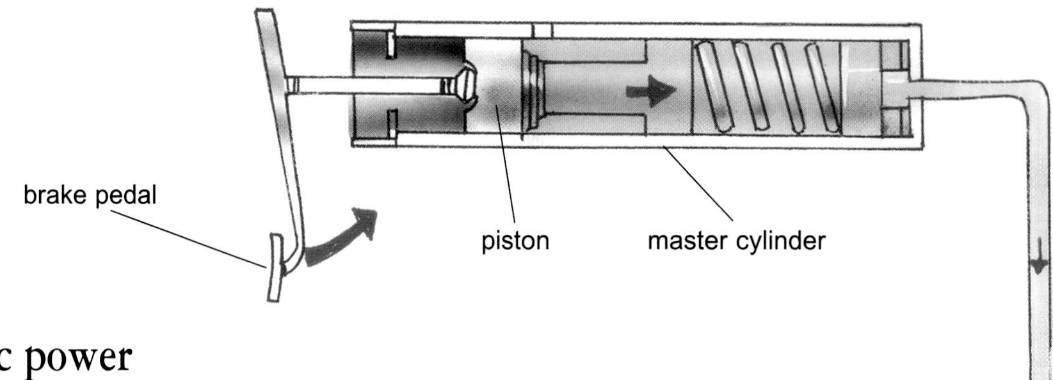

Hydraulic power

Some machines use liquid pressure to bring about movement. We call this hydraulic power. Airplanes, for example, use hydraulic power to move the rudder and the flaps and lower the landing gear.

Closer to home, a car's brakes work by hydraulic power. Look at the diagram. A driver applies the brakes by pressing down on the brake pedal. This action forces a piston along a cylinder. In turn, this forces liquid along pipes leading to the brakes on each wheel. There the liquid pressure acts against pistons. The pistons force the brake pads against the wheel disk and slow it down.

Every hydraulic power system uses liquid pressure acting against pistons to bring about movement. A unit consisting of a piston and cylinder is called a hydraulic ram.

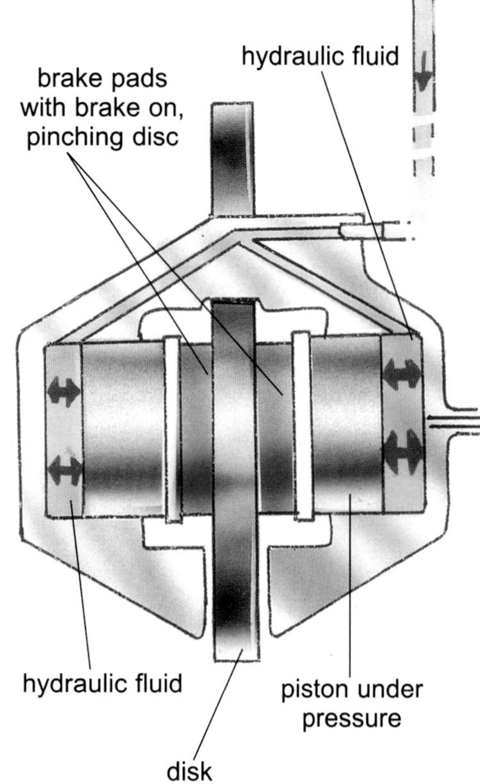

▲ Outline of how car disk brakes work. Pushing on the brake pedal creates pressure in the hydraulic fluid. The pressure is transmitted to the pistons, which force pads against the wheel disk.

◀ The lifting mechanism of this fork-lift truck works by hydraulic pressure

Electric power

One of the most useful ways of driving machines is by electric motor. Motors can be made any size and any power. They can be used to power all kinds of machines from electric toothbrushes to locomotives.

Electric motors are very compact for the power they produce. Compared with other engines, they are very clean. They produce little heat and no polluting gases. They can be used anywhere that a supply of electricity is available.

The motors are also simple in construction. They consist of a rotor and a stator, both of which are formed from coils of wire. When electricity is passed through the motor, a pattern of magnetism is set up that forces the rotor to turn.

squirrel cage motor

electromagnetic coils

▶ This is one of the commonest kinds of electric motor, called the induction motor. The rotor is made up of a series of bars, and is called a squirrel cage.

▼ Electric locomotives are driven by powerful locomotives. This type picks up electricity from overhead power lines.

Perpetual motion

Centuries ago, ingenious people tried to invent machines that would work without engines. Once these machines were set in motion, they would continue moving forever. Such machines are called perpetual motion machines.

Unfortunately, it is impossible to build a perpetual motion machine. This is because in every machine, no matter how ingenious, friction is at work, and friction absorbs energy. Eventually, it will remove all the energy in a machine. Or in other words, the machine will grind to a halt.

Nevertheless, many ideas for perpetual motion machines have been put forward. Some are illustrated here.

In this machine, the bulbs are partly filled with mercury. As a bulb reaches the top, the movement of the mercury turns the gears and rotates the wheel.

▶ **This satellite is in orbit, traveling at a speed of about 17,500 mph (28,000 km/h). It stays traveling at this speed in the same orbit for years at a time in a kind of perpetual motion.**

Q Why doesn't the satellite slow down?

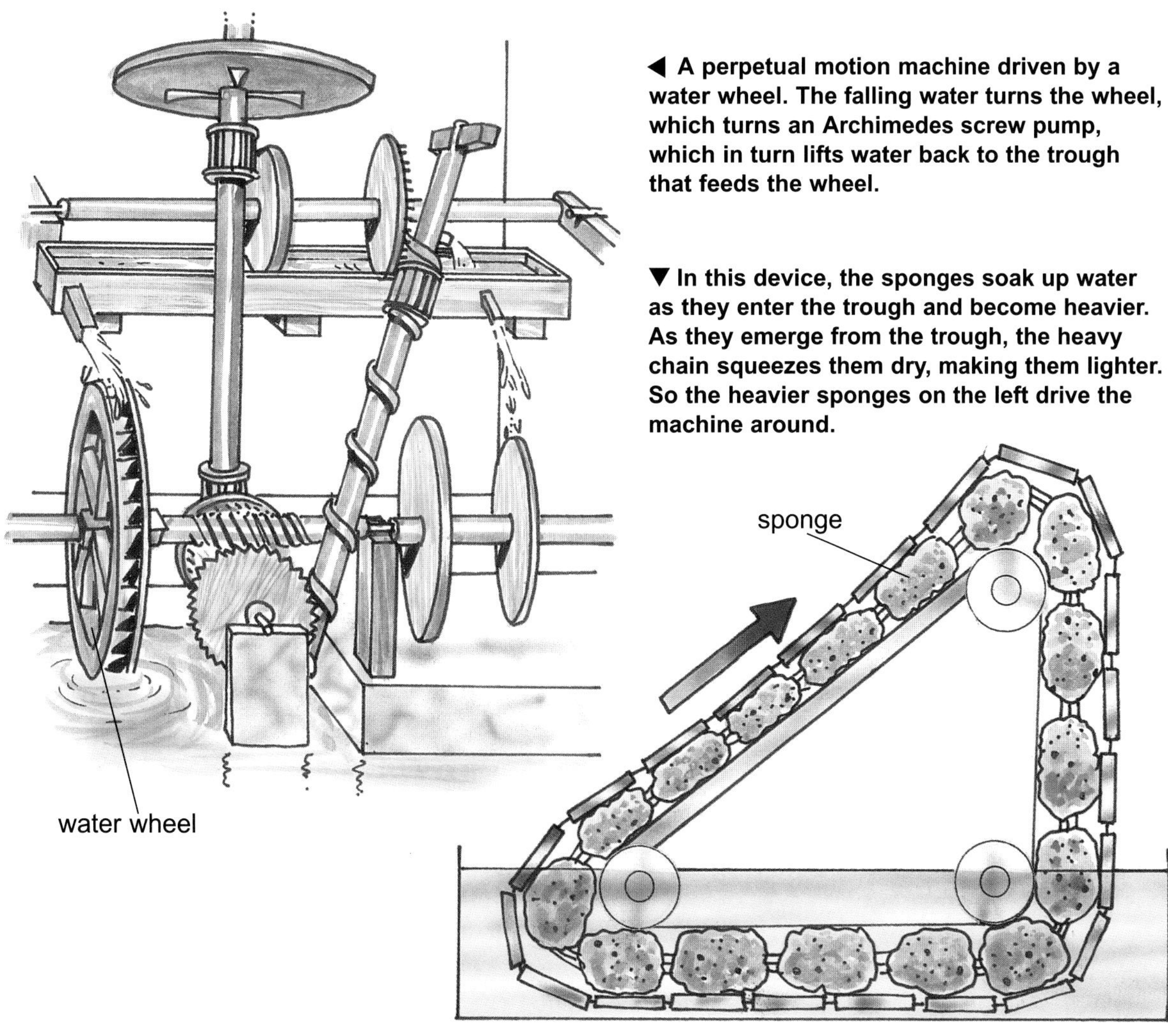

◀ **A perpetual motion machine driven by a water wheel. The falling water turns the wheel, which turns an Archimedes screw pump, which in turn lifts water back to the trough that feeds the wheel.**

▼ **In this device, the sponges soak up water as they enter the trough and become heavier. As they emerge from the trough, the heavy chain squeezes them dry, making them lighter. So the heavier sponges on the left drive the machine around.**

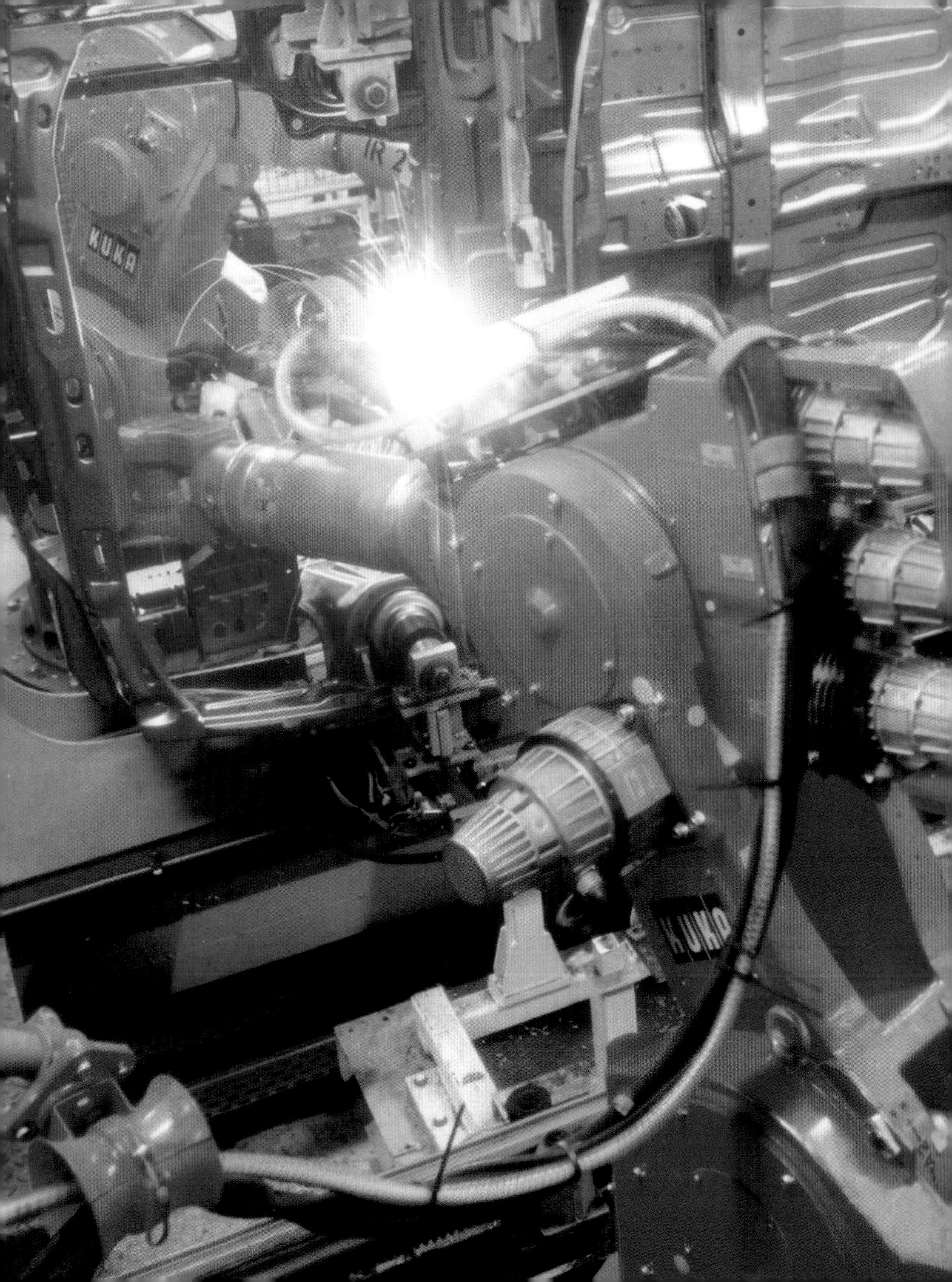
KUKA
IR 2
KUKA

3

Machines at Work

◀ **Robots at work on a car assembly line. They work alongside humans, doing repetitive and dirty tasks such as paint-spraying and welding.**

In this chapter, we look at some practical machines that we find in industry, on the roads, on the farm, and in the home.

Few machines of any kind were used until about 5,000 years ago. The wheeled wagon and the plow, drawn by oxen, were among the earliest machines. Later, the Romans built useful cranes with pulley blocks and powered by a human treadmill. They also developed the waterwheel. This was a great milestone in technology, when humans began harnessing the power of nature.

Water power remained the most important power source for driving machinery until the 1700s. That was when Scottish engineer James Watt developed an effective steam engine. At the same time, people began designing machines that could do the work of many workers and harnessed them to steam engines. This ushered in a period of history known as the Industrial Revolution. The Revolution not only changed industry; it also completely altered the ways of life of much of the population.

▶ **Machines to make textiles spearheaded the Industrial Revolution in the mid-1700's. Here, a husband and wife are using machines to make stockings.**

Textile machines

Perhaps surprisingly, the Industrial Revolution started in textile-making – the spinning and weaving of cloth. The textile industry was the first true industry.

Two main processes are involved in making cloth – spinning and weaving. In spinning, fibers are mixed together and drawn out into a long thread and twisted. In weaving, one set of threads is threaded under and over another to make the cloth.

Spinning

Spinning is necessary in making cloth because most natural fibers, such as cotton and wool, are short. They have to be gathered together into a long rope and then drawn out and twisted. This makes them into a continuous strong thread.

The diagram shows the main stages in a typical spinning process, using cotton. The fibers are first cleaned and formed into a loose blanket, or lap. This then goes to a card-

▲ The original cotton gin. Eli Whitney invented the gin in 1791 to separate the fibers from cotton seeds. It did the work of 50 people separating the fibers by hand.

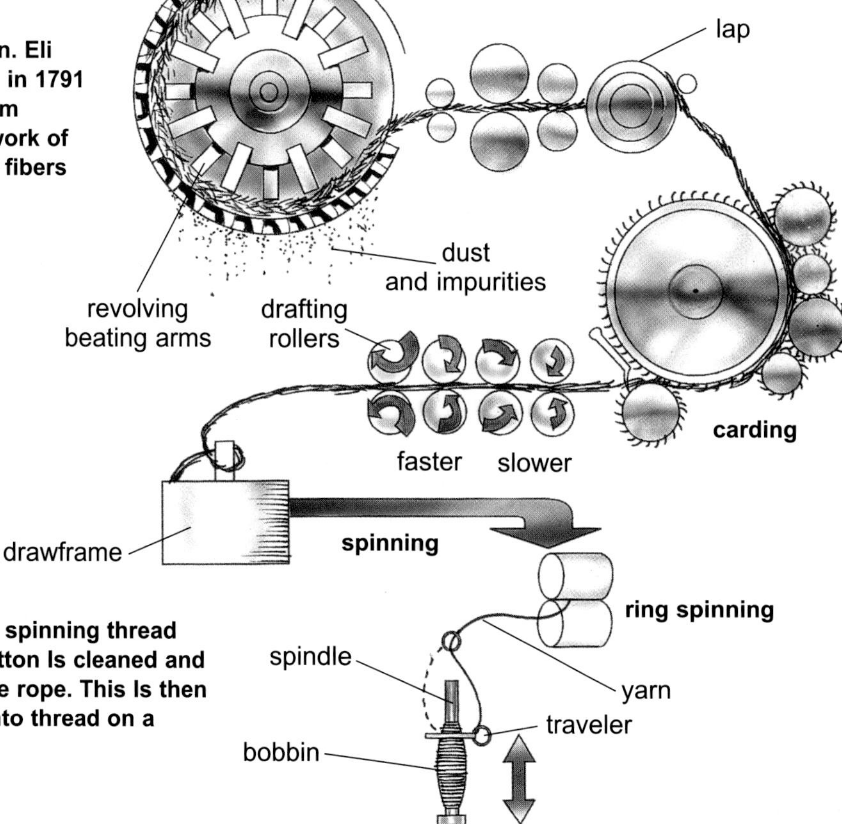

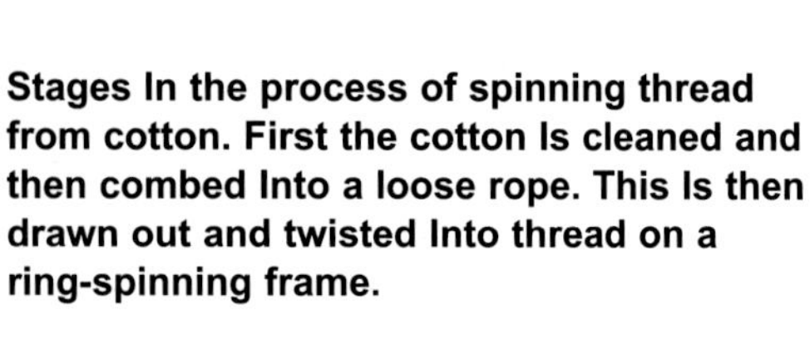

Stages In the process of spinning thread from cotton. First the cotton Is cleaned and then combed Into a loose rope. This Is then drawn out and twisted Into thread on a ring-spinning frame.

ing machine, which combs the fibers straight. The cotton comes from this machine as a loose rope. This rope is drawn out as it passes through a series of rollers traveling at increasing speeds. The now finer rope is ready for the spinning machine.

The commonest machine used to spin cotton is the ring-spinning frame. The principle of ring spinning is illustrated here. The action of the ring moving around the bobbin twists and winds the thread.

WORKOUT

If one worker can separate 1 pound (0.5 kg) of cotton fibers in a day, what weight of fibers could a cotton gin separate in a six-day work week?

Weaving

Weaving the thread into cloth is done on a loom. Industrial looms work on much the same principle as the hand loom illustrated, but they are powered and work incredibly fast. The fastest ones don't have an ordinary shuttle. They pass threads back and forth by means of jets of air or water.

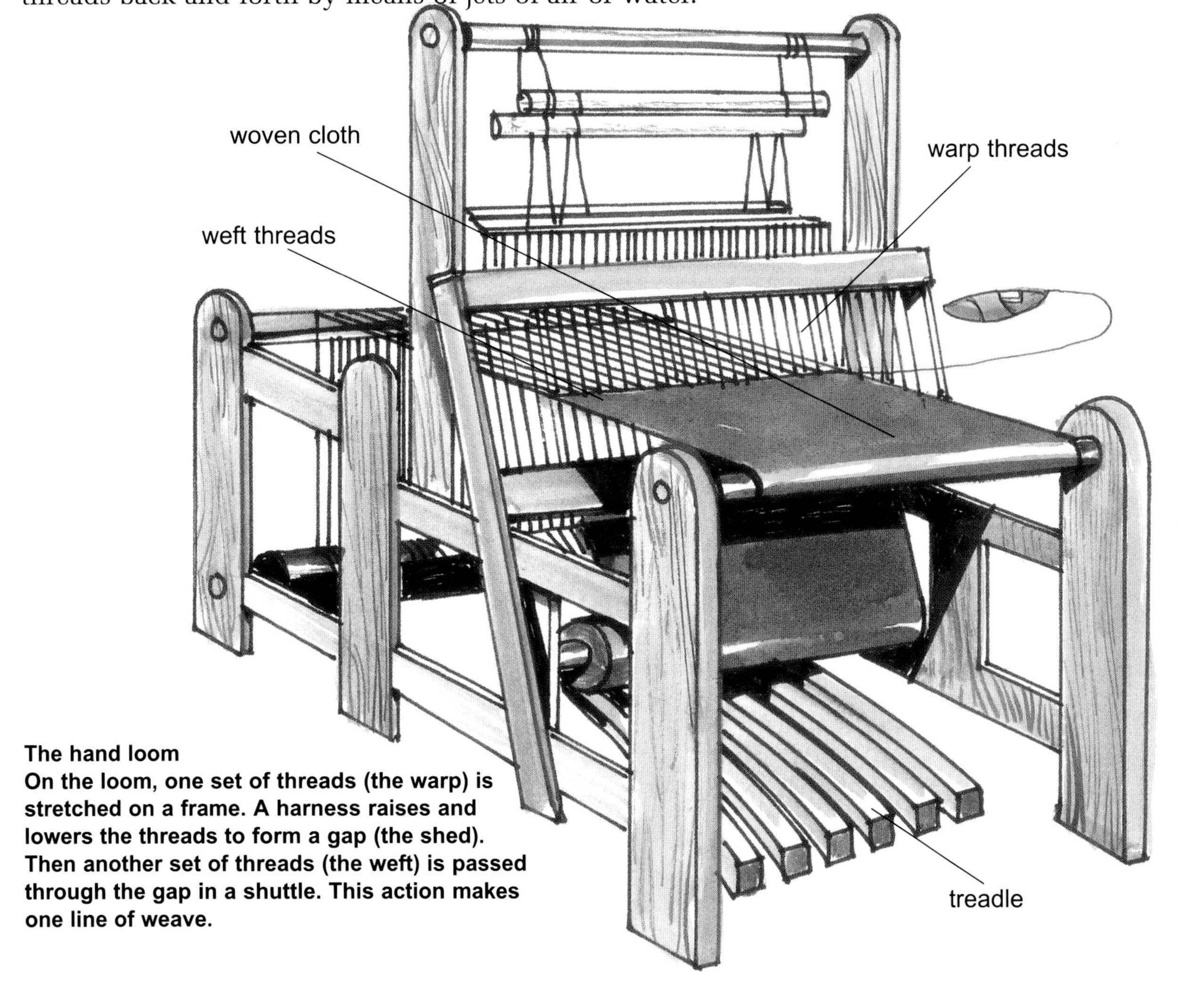

The hand loom
On the loom, one set of threads (the warp) is stretched on a frame. A harness raises and lowers the threads to form a gap (the shed). Then another set of threads (the weft) is passed through the gap in a shuttle. This action makes one line of weave.

Machine tools

The first machines that came into use in the 1700s were not very well made. For example, the cylinders for steam engines weren't bored accurately, so the pistons didn't fit properly. Steam escaped past them, which reduced power. Then a British engineer, John Wilkinson, built a machine that could bore cylinders very accurately, greatly improving steam engines.

This accurate boring machine was the first precision machine tool. Machine tools are machines that cut and shape metal very accurately. They hold the key to the construction of all other machines, from typewriters to cars.

Machining operations

Machine tools carry out a number of basic metal-cutting, or machining operations. They are illustrated below. Drilling cuts holes in metal with a rotating bit. The machine tool for drilling is called a drill press. Often drill bits of different sizes are mounted on a revolving turret.

Grinding is done by a rotating wheel covered with a rough substance, or abrasive. Shaping uses a chisel-like blade to remove metal. Milling uses a rotating wheel with cutting teeth. The wheel may cut vertically or horizontally. Turning is a machining operation in which a cutting tool removes metal from an object that is rotating.

▼ The lathe Is one of the most common machine tools found in factory workshops. It carries out the process of turning. A piece of metal is held between the headstock and tallstock and rotated. Tools are moved in to cut the metal as it rotates.

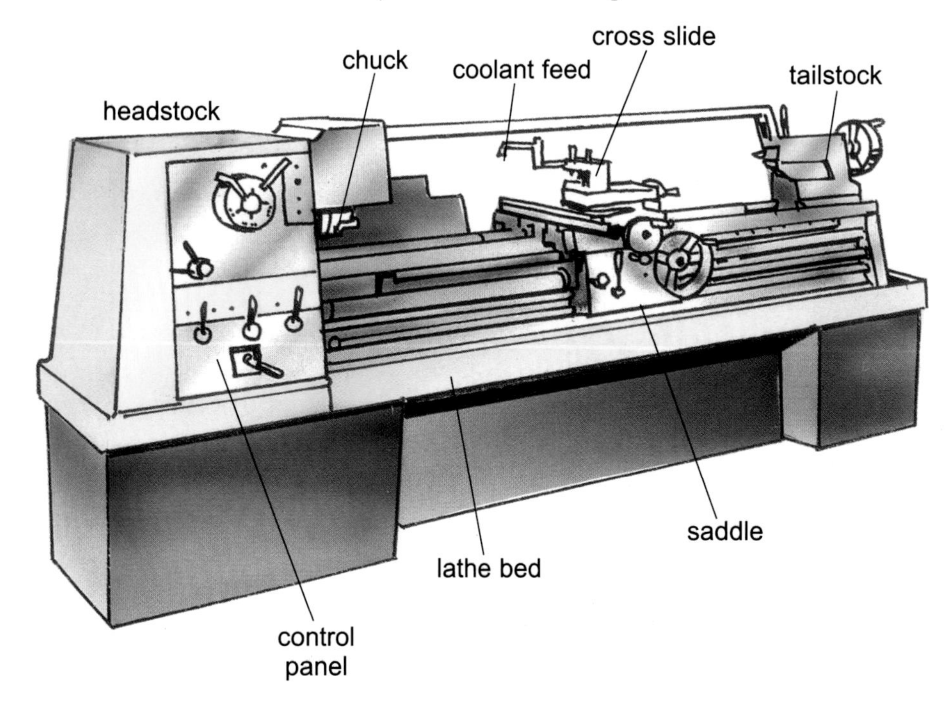

1.

2.

3.

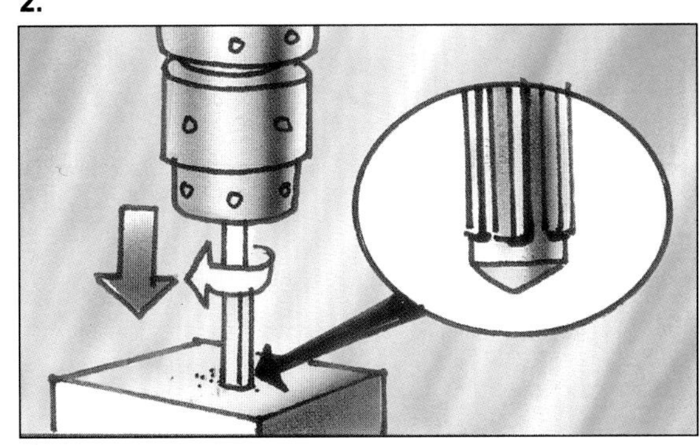

4.

5.

6.

Common machining processes carried out by machine tools Include (1) grinding, (2) turning, (3) drilling, (4) reaming, (5) shaping, and (6) milling.

Power machines

Machine tools are designed to cut through metal, which is, of course, hard and strong. They are driven by powerful electric motors. Usually they include a gearbox, which allows them to run at different speeds for different jobs.

The cutting tools are made of specially hardened steel, which can cut through ordinary metal. For high-speed operation, the tools are made of special alloy steels. These contain heat-resistant metals such as tungsten.

Q In most machine tools, a liquid called a cutting oil is fed into the cutting area. Why do you think this is done?

Putting machines together

Most of the objects we use are not made all in one piece. They are made by assembling, or joining together, a number of smaller parts, or components.

As you can see from the cutaway picture here, a car is a very complicated machine. It is put together from up to 15,000 separate parts, from nuts and bolts and ball bearings to engine blocks and steel body panels.

These parts are put together on an assembly line.

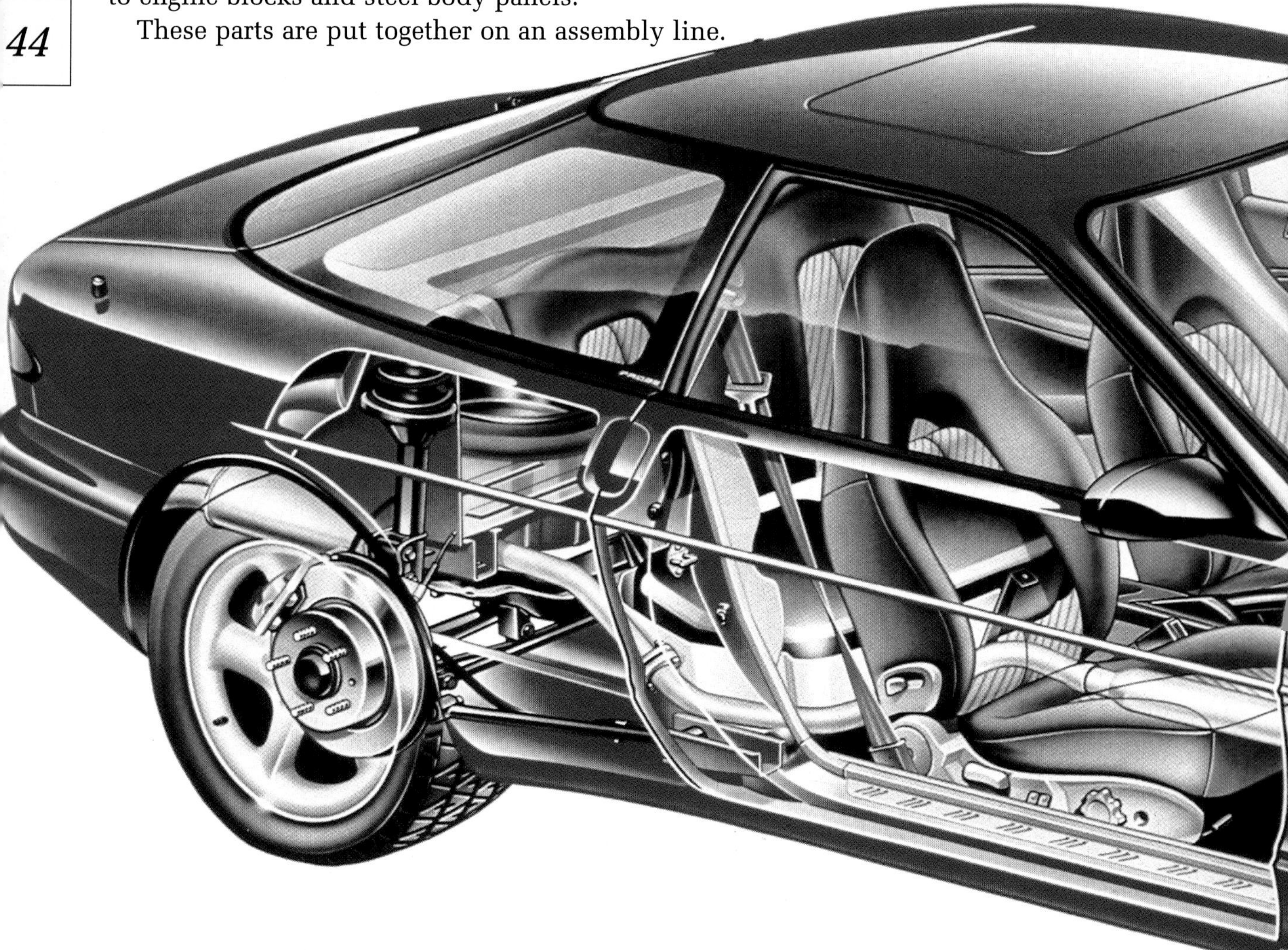

Above: A modern car is the most complicated machine we come across in our everyday lives. On a car production line, human workers and robots work together to assemble the machines from sets of identical parts.

Opposite top: On the production line, robots carry out their alloted tasks with unerring precision.

Workers stand in line, with certain parts on hand, as the skeleton of the car moves past on a conveyor. They add parts to the skeleton, which then moves on so that other workers can add more parts. At the end of the line, the car is finished, ready to drive away.

This method of manufacturing is very efficient, but it is only possible because of machine tools. That is because they are able to turn out virtually identical parts every time.

Q **1.** What would happen on an assembly line if the parts workers used weren't identical?

Q **2.** Who first used a moving assembly line to build cars?

He also produced only one famous model. What was it called?

Car mechanics

The car is by far the most complicated machine we come across in our everyday lives. As mentioned earlier, we can think of it as two machines in one. One is the engine, which provides turning power. The other is the transmission, which transmits the turning power to the driving wheels. We dealt with the basic mechanics of the car engine in the last chapter. Here we deal with the basic mechanics of the transmission. It consists mainly of shafts linked by gears.

The transmission system illustrated here is for a car that has a front engine driving the rear wheels. The engine motion is transmitted to the wheels through a clutch, gearbox, propeller shaft, and final drive. The clutch acts like a switch to make or break the connection between the engine and the gearbox. The gearbox changes gears to make the engine drive the wheels at different speeds. The propeller shaft carries the motion from the gearbox to the final drive. The gears in the final drive change the speed and direction of motion to drive the wheels.

Most American cars have an automatic transmission . The car here is shown with a manual or stick-shift transmission, because this type shows the gears better.

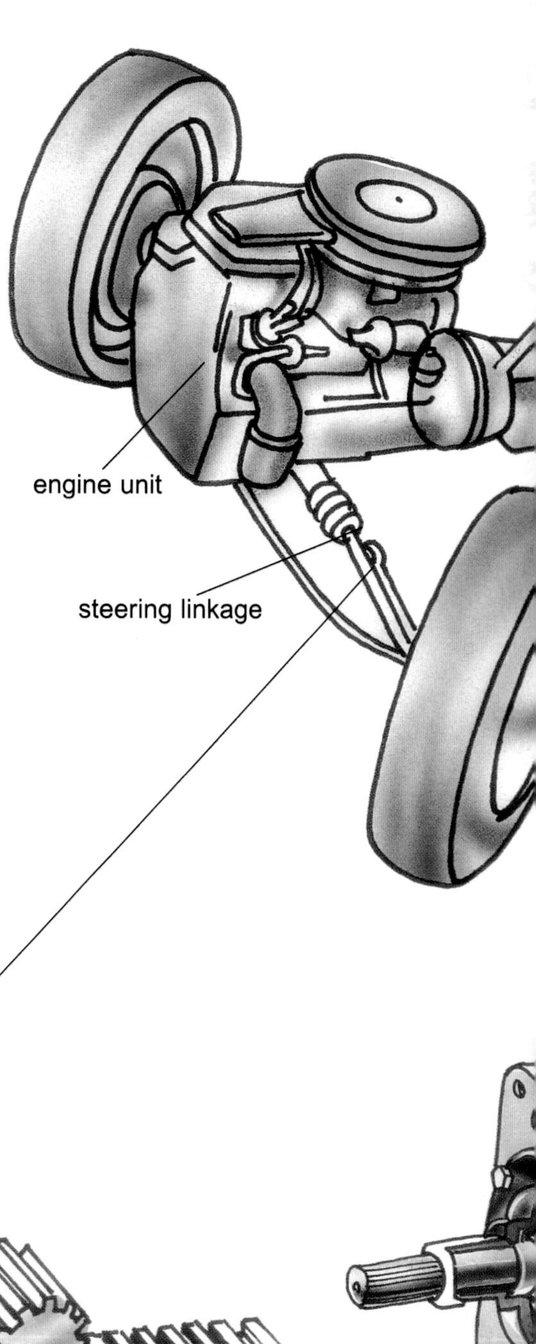

▶ **The steering wheel moves the front wheels of the car by means of a rack-and-pinion gear. A toothed gear, or pinion, on the end of the steering wheel shaft meshes with a toothed rack that connects with the front wheels. When the steering wheel is turned, the pinion moves the rack from side to side.**

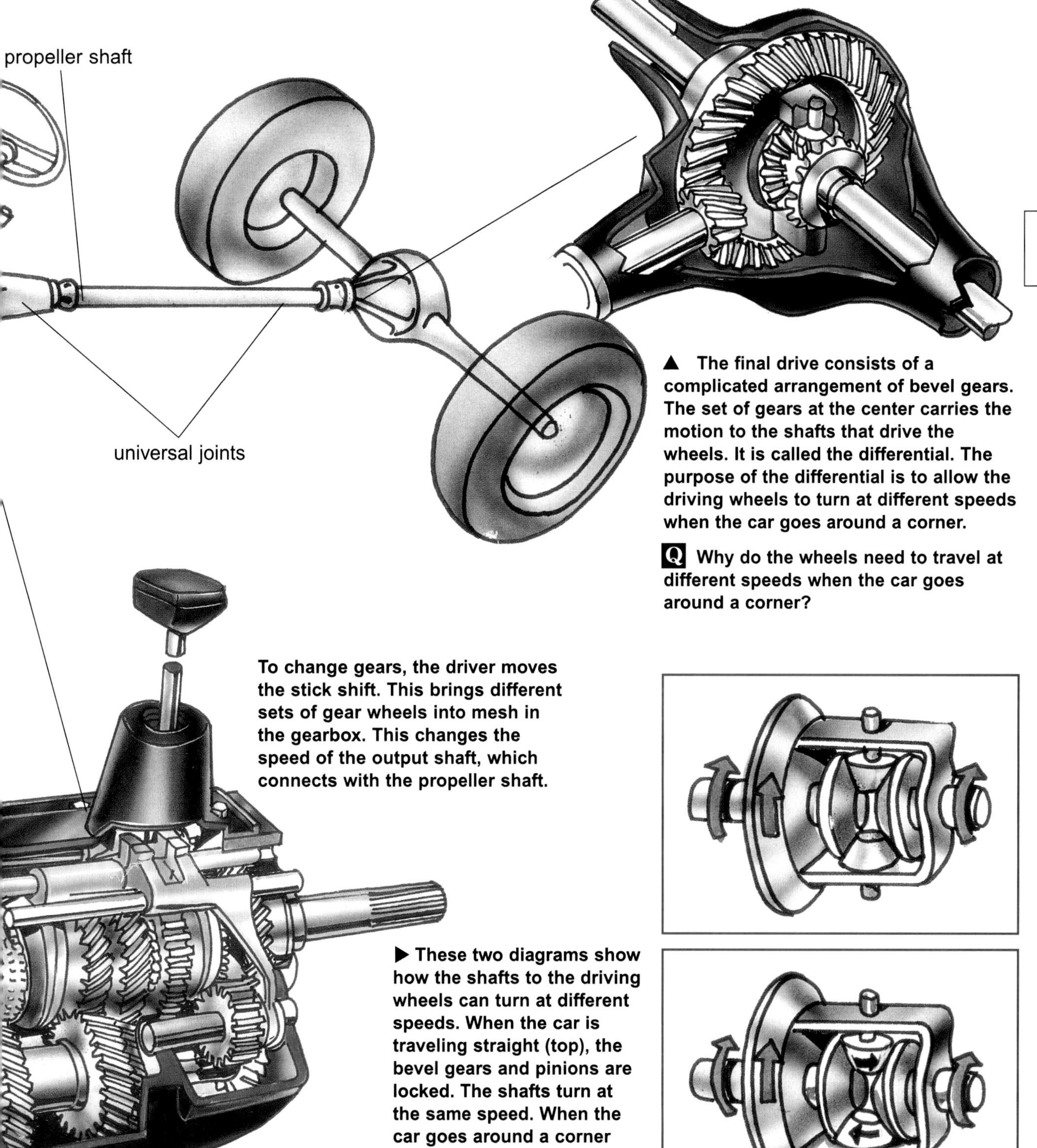

▲ The final drive consists of a complicated arrangement of bevel gears. The set of gears at the center carries the motion to the shafts that drive the wheels. It is called the differential. The purpose of the differential is to allow the driving wheels to turn at different speeds when the car goes around a corner.

Q Why do the wheels need to travel at different speeds when the car goes around a corner?

To change gears, the driver moves the stick shift. This brings different sets of gear wheels into mesh in the gearbox. This changes the speed of the output shaft, which connects with the propeller shaft.

▶ These two diagrams show how the shafts to the driving wheels can turn at different speeds. When the car is traveling straight (top), the bevel gears and pinions are locked. The shafts turn at the same speed. When the car goes around a corner (bottom), the pinions turn, allowing the shafts to turn at different speeds.

Lifting machines

From the earliest times, human beings have been builders, not only of low-rise houses but also of magnificent structures and monuments, such as Stonehenge and the pyramids. They used poles and ropes to help them lift and maneuver huge blocks of stone into place. By 2,000 years ago, the Romans were using lifting devices consisting poles, ropes, pulleys, and a winch, or winding mechanism. They made up an early kind of crane.

As a machine, the crane combines the lifting power of the pulley with the pulling power of a winch. A winch is a kind of wheel and axle, one of the simple machines.

Today, cranes are in widely used on building sites, in factories, and in dockyards. They all have certain things in common. The load is lifted by a hook, which is attached to a pulley block. The pulley block can be moved up or down by winding a wire cable, or rope, on a winch. In most cranes, the block hangs from a long arm called the boom or jib.

Kinds of cranes

The most familiar crane on building sites is the tower crane. It has a horizontal jib, which swings around on top of a tower. The pulley block hangs from a trolley that travels along the jib.

◀ The tower crane is a familiar sight on construction sites. It can cover a wide area.

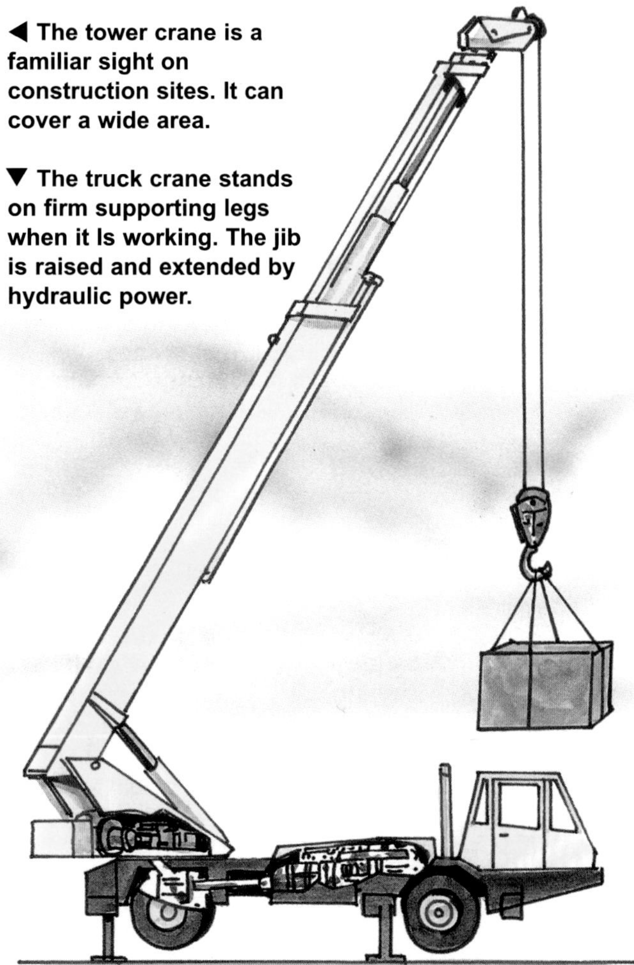

▼ The truck crane stands on firm supporting legs when it Is working. The jib is raised and extended by hydraulic power.

A mobile crane has a jib that can swing around and move up and down. It may be mounted on caterpillar (crawler) tracks or on a truck body. Truck cranes have jibs that extend like a telescope.

Elevators and escalators

Cranes are designed for lifting things, but machines have also been designed to lift people. The two most important are the elevator and the escalator, or moving staircase. American inventor Elisha Otis demonstrated the first safe elevator in 1853. The first escalator was built for the Paris Exposition of 1900.

An elevator uses a simple mechanism. The passenger car hangs from wire cables that go up and over a pulley wheel at the top of the elevator shaft. A heavy counterweight hangs from the other end of the cables. The pulley wheel is driven around by a powerful electric motor. This moves the cables, causing the passenger car to rise or descend.

Q Why do you think that this arrangement is better than using a simple winch to move the car up and down?

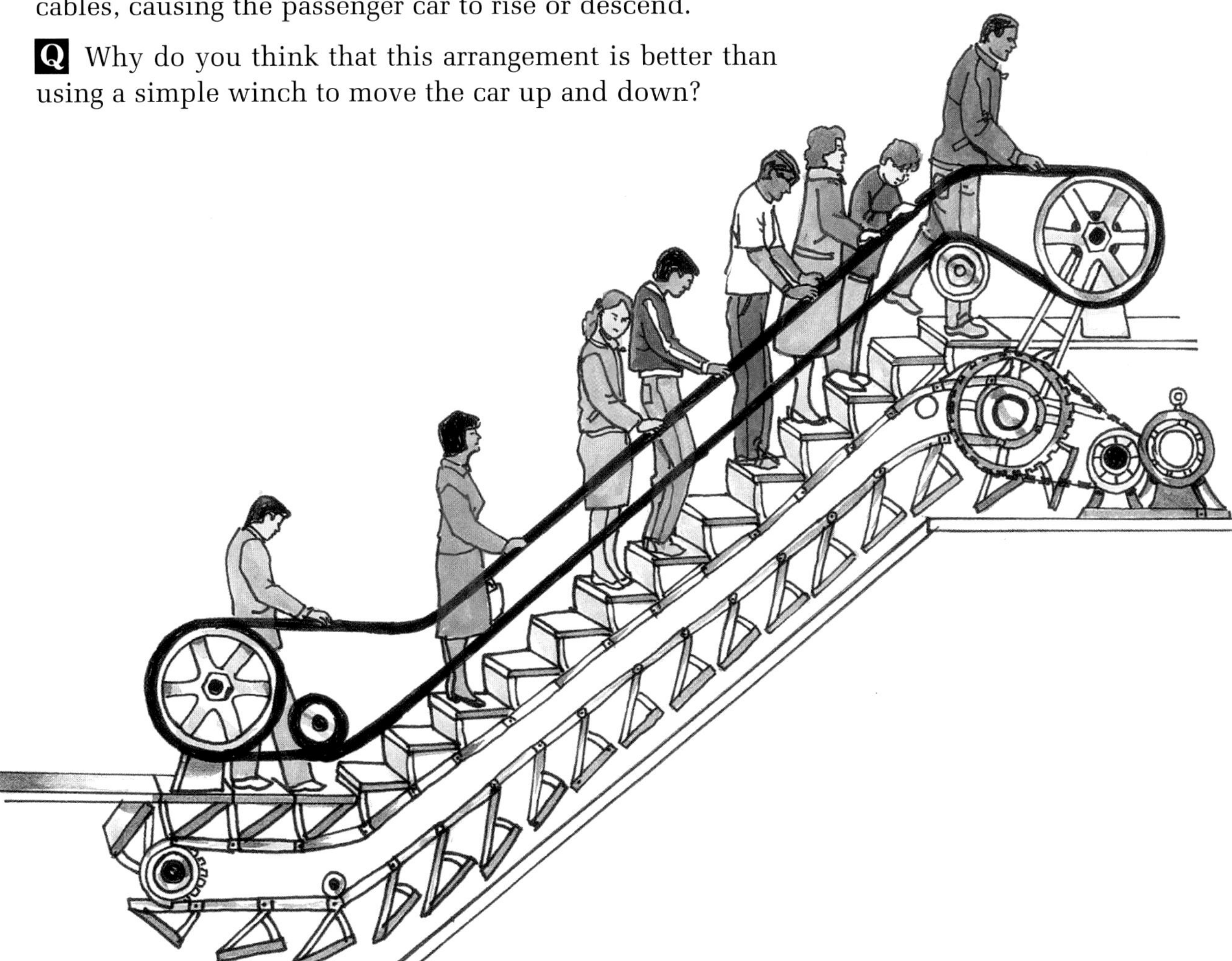

Digging and earthmoving

Another common machine on building sites is the bulldozer. This machine is one of the first machines on many sites because it is used to clear the ground before construction. It has a tough steel blade in front to move soil, rubble, or tree stumps. Bulldozers move on caterpillar tracks, which are endless chains of steel plates.

Q Why do you think bulldozers have caterpillar tracks rather than wheels?

▲ A powerful bulldozer uses its steel blade to clear rubble from a construction site.

▶ This machine has a host of uses in building construction. It can be used for digging, bulldozing, and lifting. It can also be fitted with a pneumatic drill to break up materials such as concrete and tarmac.

Earth-moving

Other earth-moving machines can be seen on larger construction projects, such as road building, One is called a scraper. This machine consists of an engine unit pulling a huge bowl with a cutting blade and opening at the front lower edge. When it moves across the ground, the blade slices into the soil, which is then forced into the bowl.

Excavators

There are several other kinds of machines fitted with digging attachments for removing soil. They are called excavators. Some are like tractors with digging buckets back and front. The digging arms are moved by means of hydraulic rams.

Larger excavators called draglines look like cranes. They fling out a digging bucket on the end of a rope and then drag it back over the ground. As it is dragged back, it scoops up the soil. The largest draglines are found at open-pit coal mines. They are used for removing the soil on top of the coal deposits.

▲ This kind of excavator is used In open-pit mining operations. It Is known as a bucket-wheel excavator. It has digging buckets arranged around a wheel that rotates.

IT'S AMAZING!

Big Muskie

A dragline excavator known as Big Muskie is working at the Central Ohio Coal Company's Muskingum site in Ohio. It is the world's largest excavator.

❊ Its boom measures 310 feet (94 meters) long.

❊ Its digging bucket can remove 220 square yards (170 square meters) of soil at one bite.

❊ It weighs 12,000 tons (11,000 tonnes).

WORKOUT

If a Boeing 747 jumbo jet weighs 360 tons (330 tonnes), how many jumbo jets could balance Big Muskie on a pair of scales?

On the farm

About a century ago, close to two-thirds of the population of the United States were farmers. Today, only about three and a half million Americans work on farms, yet these few farmers produce more than enough food for the rest of the population.

One of the main reasons why so few farmers can produce so much food is because of mechanization – the use of machines. The use of fertilizers and pesticides have also played a part.

Q How do fertilizers and pesticides help to increase food production?

Powerful tractors are the most important machine on the farm. They have to be powerful in order to pull the latest implements, such as the 10-bottom plow (below left).

The tractor

The most common machine on any modern farm is the tractor, which has taken over from the horse as the "farmer's best friend." This tough, rugged vehicle is built for pulling power rather than speed. It has huge wheels at the rear with a thick tread to give a good grip on the ground.

The powerful engine usually drives the four wheels rather than just two to give extra pulling power. It also drives a shaft called the power take-off. This is used to drive implements that the tractor pulls, such as a mower. The tractor is also fitted with a hydraulic power mechanism for raising and lowering implements such as plows.

Plows are implements that have a set of sharp blades that slice into and turn over the soil. Plowing is the first stage in preparing the ground for the planting of a new crop. After plowing, farmers break up the plow furrows with a harrow. Then they sow seeds with a seed drill. This machine delivers the seed at the right depth beneath the soil. Later, as the crop starts growing, farmers spray it with chemicals to ward off diseases, such as mold.

WORKOUT

Find out from your library what the population of the United States is at present. Using the number of farm workers given in the text, work out the percentage of people that are farmers. Also, express this information as a pie chart.

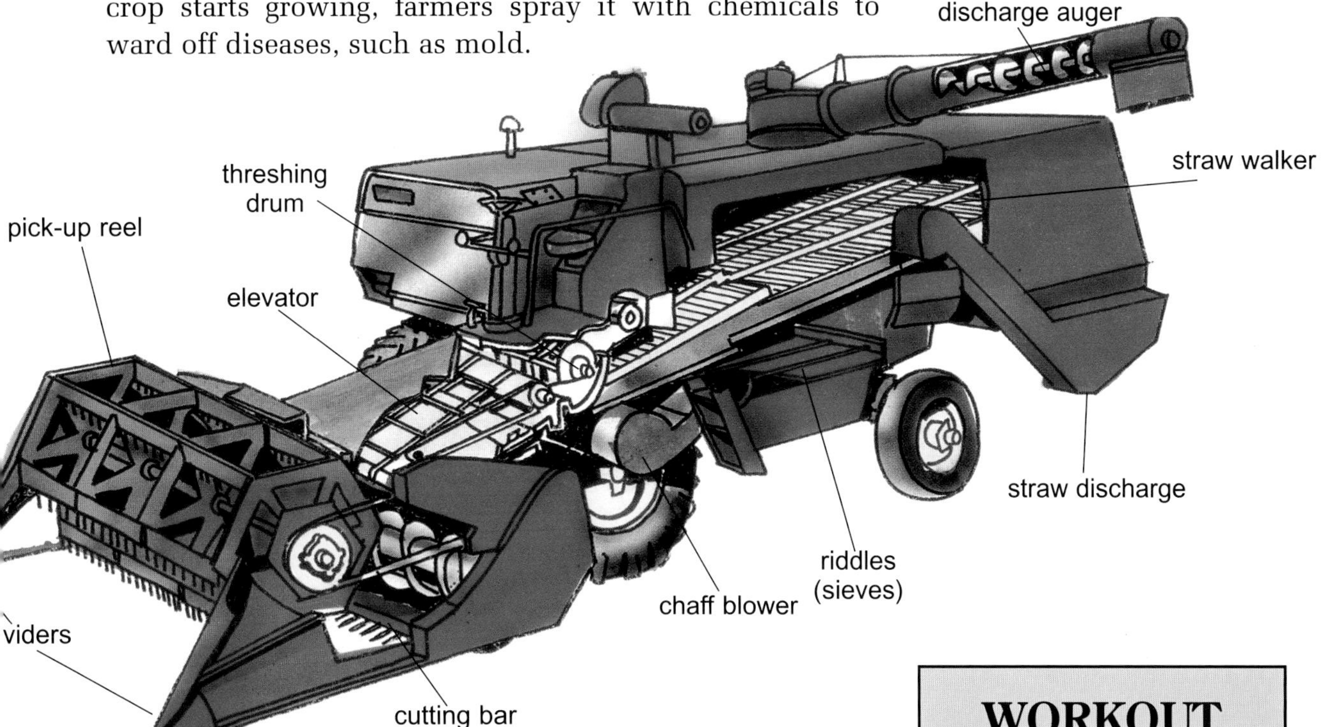

Harvesting

Farmers use a variety of machines for harvesting their crops. By far the most useful is the combine harvester, which is used mainly to harvest cereal crops, such as wheat and barley.

The combine is so called because it combines the actions of reaping and threshing. Reaping means cutting the crop, and threshing means beating the crop to force the grains out of the ears.

WORKOUT

Some combine harvesters have a cutting width of 30 feet (9 meters). Using such a machine, a worker can harvest 10 acres (4 hectares) of land per hour. How long would it take four workers using such machines to harvest an area of 200 acres?

Word processing helps speed up basic office operations, such as letter-writing and filing. Mostly, workers use computers with a word-processing program. They may also use the computer for other purposes, such as processing data or sending faxes or electronic mail.

Machines in the office and home

As we have seen, machines speed work in the factory, in construction, and on the farm. Machines also speed work in the office and the home.

The major machine that brought greater efficiency to the office was the typewriter. This came into widespread use near the end of the 19th century. The typewriter works by means of a series of pivoted levers. When you strike a key on the keyboard, a lever called the type bar springs forward. It presses a piece of type against an inked ribbon, which makes contact with a sheet of paper. Many modern typewriters work electrically. But in most offices today, typewriters are being replaced with word processors, which are computer-based type-writing systems.

Photocopiers

The photocopier is another machine found in almost every office. It is often called a xerox machine. This is because it uses a process known as xerography to make copies. The word means "dry writing," referring to the fact that it doesn't use wet ink to make copies.

The diagram below shows the essential parts of a photocopier. The page to be copied is placed, printed side down, on the glass at the top. A bright light scans over the page. An image of the page is reflected onto a specially coated drum, which is charged with electricity. Where the image is light (no ink), the electric charge leaks away, but where the image is dark (inked), the electrical charge remains.

These charged areas now attract a black powder, called toner. When a sheet of paper is pressed against the drum, the toner transfers to it. A heated roller melts the toner to form a permanent image on the paper.

Q What is the name of the science that deals with electric charges?

A look inside a typical office photocopier. The coated drum records an image of the document to be copied. Special powder Is attracted to the image and then transferred to paper.

lamp

heated rollers

drum

mirror

paper sheet

Home helps

Machines to make housework easier began to appear during the 19th century. One of the first was the sewing machine. Modern machines developed from the successful machine produced by Isaac Singer in 1851. The basic stitch of the machine is the lock stitch. It is formed using two sets of threads, one from a needle and one from a bobbin. The diagram below shows how the lock stitch is formed.

In 1908, a leather manufacturer named William Hoover sold his first machine for cleaning carpets. So was launched the vacuum cleaner, or "hoover." It is a relatively simple device, in which a rotating cylinder beats and sweeps the carpet and a fan sucks up the dust into a bag. Says Hoover's famous advertising slogan: "It beats as it sweeps as it cleans."

suction fan

brush

beater bar

Above right: Mechanically, the vacuum cleaner is quite simple. The fan and beater bar are driven by an electric motor.

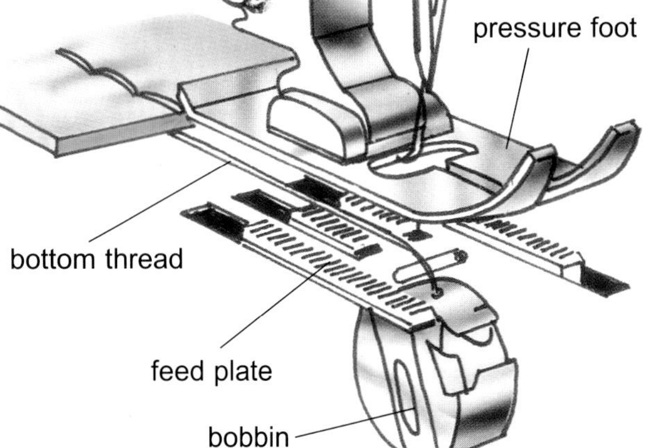

Right: A rotating bobbin holds the key to mechanical sewing. It carries the lower thread for stitching, while the needle carries the upper thread. The sequence below (1-4) shows how the threads loop to form a lock stitch as the bobbin rotates.

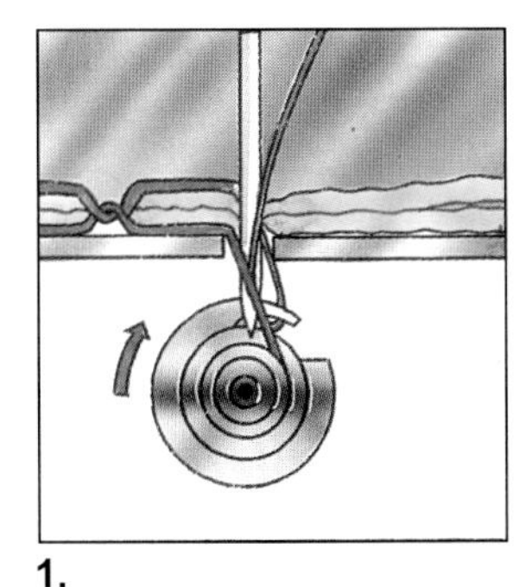

1.

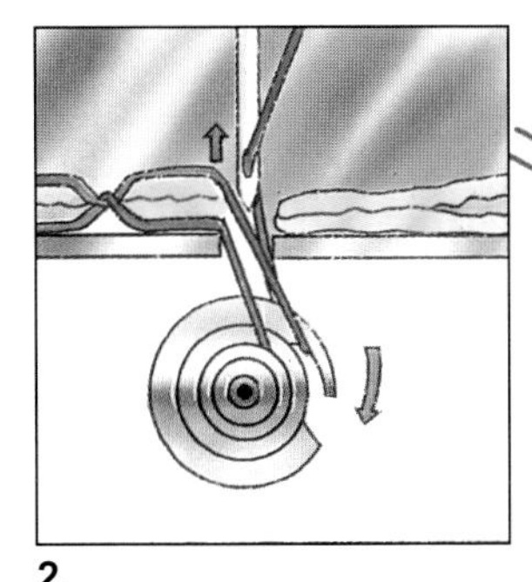

2.

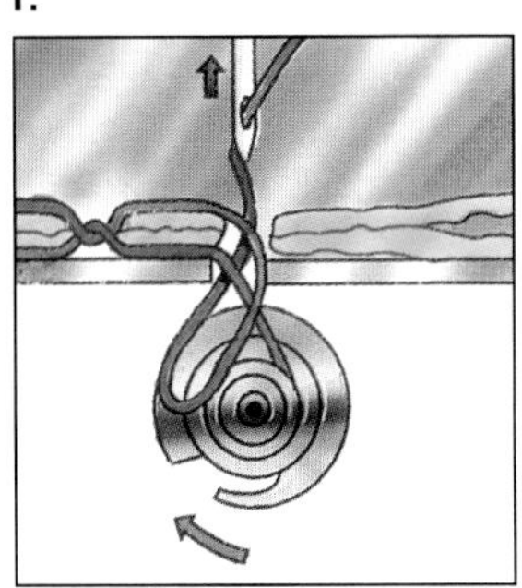

3.

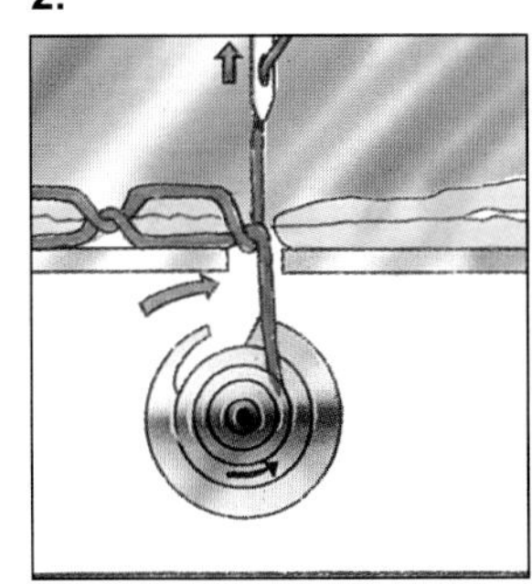

4.

Milestones

ca. 250,000 BC Stone axes were first produced.

ca. 3500 BC The potter's wheel came into use in the Middle East. Somewhat later, the wheel began to be used on wagons for transportation.

ca. 400 BC Archytas, from southern Italy, introduced the pulley,

ca. 240 BC The Greek scientist Archimedes discovered the laws of the lever and invented the Archimedean screw for lifting water.

ca. 100 BC The Romans began using watermills to grind grain for making flour.

ca. 10 BC Roman engineers used cranes for lifting things.

ca. AD 250 The Chinese began using wheelbarrows.

ca. 650 Windmills came into use in Persia, now Iran.

1300s Mechanical clocks were introduced in Europe.

1511 Peter Henlein in Germany made the first watch.

1712 Thomas Newcomen in Britain developed a piston steam engine for pumping.

1767 James Hargreaves in Britain built the spinning jenny, a machine to speed up spinning.

1792 Eli Whitney introduced the cotton gin in the United States.

1803 Richard Trevithick in Britain built thc first steam locomotive.

1837 Thomas Davenport of Virginia demonstrated a practical electric motor.

1839 Scot Kirkpatrick Macmillan built the forerunner of the modern bicycle.

1858 Isaac Singer in the United States introduced the domestic sewing machine.

1862 Richard Gatling invented the Gatling gun, the first machine gun. Fellow American L.O. Colvin developed a practical milking machine.

1869 George Westinghouse developed air brakes for railroad trains.

1872 Christopher Sholes invented a practical typewriter.

1884 Charles Parsons in Britain built the first successful steam turbine.

1885 Gottlieb Daimler in Germany built the first motorbike. Fellow countryman Karl Benz built the first successful motor car.

1889 Charles Fey of San Francisco built the first fruit machine, the "Liberty Bell."

1892 Rudolf Diesel in Germany developed the diesel compression-ignition engine.

1898 The Huber Co. of Marion, Ohio, began to build practical tractors for farming use.

1901 Hubert Booth In London invented the vacuum cleaner. William Hoover began manufacturing a portable domestic model seven years later.

1905 John Danton of Michigan built the first modern jukebox.

1913 Henry Ford in the United States introduced a moving assembly line for mass-producing cars.

1923 The US LaPlante-Choate company Introduced the bulldozer.

1930 Frank Whittle of Britain patented a jet engine.

1936 Heinrich Focke designed a practical helicopter.

1946 The Ford Motor Co. plant at Detroit Introduced automation, with the widespread use of automatic machines.

1967 First successful flight of the Saturn V Moon rocket, designed by Wernher von Braun.

1981 Maiden flight of the space shuttle, part rocket, part spacecraft, part glider.

1983 The jet-powered car Thrust 2 became the fastest land machine when it reached a speed of 633.5 mph (1,019.5 km/h) over the Black Rock Desert In Nevada.

Glossary

ASSEMBLY LINE A factory production system in which workers stand In line and assemble a product bit by bit as it travels slowly past them on a conveyor.

AUTOMATION The widespread use of machines that work automatically under computer control.

AXLE A shaft on which a wheel turns.

BALL BEARING A kind of bearing in which a cage of rolling balls supports a rotating shaft.

BEARING A machine component that supports a shaft and allows It to turn with the minimum of friction.

BLOCK AND TACKLE A unit made up of several pulleys, linked together by ropes ("tackle").

BOLT A fastening device with a hexagonal head, a cylindrical body and a screw thread. A nut screws on to it.

BRAKE A device that slows something down. Vehicles are fitted with brakes to control their speed.

BULLDOZER A heavy tractor with a tough steel blade in front, used for clearing construction sites, for example. It is fitted with caterpillar tracks.

CAMSHAFT A rotating shaft carrying raised projections, called cams. It is used in engines to open and close valves, for example.

CATERPILLAR TRACK Also called a crawler track. A track consisting of an endless belt of steel plates. Tractors and other vehicles are often fitted with these tracks to give them better grip on the ground. They are often called track-laying vehicles.

COG Another term for a gearwheel.

COMBINE One of the most common farming machines, used for harvesting grain and other crops. It both cuts the crop and separates the seeds. It thus combines the actions of reaping and threshing.

COOLANT A liquid used to cool and lubricate the cutting surfaces in machine tools. It is often called a cutting fluid.

CRANE A common lifting machine found on construction sites.It carries loads on the end of a rope (cable) that passes up and over a long jib (arm). It has a power driven winch for winding in the rope.

CRANKSHAFT A rotating shaft in a piston engine that changes the up-and-down movements of the pistons into rotary (turning) motion.

DIESEL ENGINE An internal combustion engine that uses a light oil as fuel. The oil burns when it is injected Into very hot air in the engine cylinders.

DISK BRAKE A kind of car brake in which the braking action forces pads against disks attached to the wheels.

DRILLING A machining operation in which holes are drilled In a workpiece with a rotating drill bit.

EFFORT The force applied to a machine.

ELECTRIC MOTOR A motor that works by electricity.

ELEVATOR A device for lifting goods and passengers.

ENGINE A machine that harnesses energy to do work. Usually, it produces motion to drive other machines.

ESCALATOR A moving staircase, widely used In stores, subways, and so on, to move passengers between different floor levels.

EXCAVATOR A digging machine. Excavators are widely used on construction sites and for digging out minerals in mining operations.

FACTORY A place in which goods are made, usually with the help of machines.

FIRST-CLASS LEVER A kind of lever In which the effort and load are on opposite sides of the fulcrum.

FOUR-STROKE CYCLE The cycle, or series of repeated operations on which many gasoline and diesel engines work. A "stroke" is an up or down movement of a piston.

FRICTION A force that affects moving bodies and opposes their motion. It is set up when one surface rubs against another.

FULCRUM The supporting point for a lever. Also called the pivot.

GASOLINE ENGINE An engine that burns gasoline as fuel. It is a piston engine that operates on a four-stroke cycle.

GAS TURBINE A turbine that uses hot gases to spin the turbine wheels. Aircraft jet engines are kinds of gas turbines.

GEARS Toothed wheels that transmit motion. By meshing together gears with different numbers of teeth, you can reduce or increase the speed of motion. This is what happens in a gearbox.

HYDRAULIC RAMS Devices that work by hydraulic (liquid) pressure. Many cranes, tractors and excavators have arms that are moved by means of hydraulic rams.

INCLINED PLANE One of the simple machines, which consists essentially of a slope.

INTERNAL COMBUSTION ENGINE (ICE) An engine In which fuel is burned Inside an enclosed space. A gasoline engine Is an ICE in which gasoline Is burned inside the engine cylinders.

JACK A lifting device. A screw jack uses the mechanical advantage of the screw to lift heavy loads.

JET ENGINE An engine that produces a stream, or jet, of hot gases. As these gases shoot backward out of a nozzle, they set up a force (thrust) forward. Most aircraft use jet engines for propulsion.

LATHE A common machine tool in which the operation of turning is carried out. A workpiece is rotated while a cutting tool moves in to remove metal.

LEVER One of the simple machines. It consists essentially of a beam, supported at one point (the fulcrum or pivot). Force or effort is applied in one place to move a weight or load at another.

LOAD A force that acts on a machine. Typically, effort is applied to the machine to overcome that force.

LOCOMOTIVE An engine that hauls a train. Most locomotives are now powered by diesel engines or electric motors.

LOOM The machine on which cloth is made by weaving.

LUBRICATION Providing oil or another fluid to overcome the friction between moving parts of a machine.

MACHINE TOOL A power-driven cutting machine used to shape metal, such as a lathe.

MACHINING A cutting operation carried out by a machine tool, such as drilling, milling and turning.

MASS PRODUCTION The production of goods on a very large scale. It is achieved In most factories by automation.

MECHANICAL ADVANTAGE Many machines exert more force than the force (effort) put into them. The mechanical advantage is the ratio of these two forces.

MILLING A machining operation in which metal is removed by a rotating cutting wheel.

NUT A fastening device with a screw thread that screws on to a bolt.

PERPETUAL MOTION Motion without end. For centuries people have tried to invent machines that, once set in motion, will never stop. This isn't possible, because of friction.

PHOTOCOPIER A machine that makes copies of documents, photographs, and so on, using a lIghtsensitive drum and special dry "Ink" or toner.

PISTON ENGINE An engine in which a piston moves back and forth along a cylinder. Diesel, gasoline, and steam engines are all piston engines.

PULLEY A simple machine, consisting of a small wheel and axle and a rope. Several pulleys are generally used together in a block and tackle.

ROBOT A machine that works automatically under computer control. An android is a robot built to resemble a human being.

ROCKET A powerful motor that burns fuel to produce a stream of hot gases. The gas stream Is used to propel space vehicles, for example. Unlike jet engines, rockets carry both fuel and the oxygen to burn it.

SCREW One of the simple machines. It has a high mechanical advantage because it moves forward only a little when turned.

SECOND-CLASS LEVER A kind of lever in which the fulcrum is at one end, the effort is at the other, and the load is in between.

SPINNING An early stage of textile-making in which short fibers are drawn out and twisted to make a strong thread, or yarn.

STEAM ENGINE An engine that uses the power of expanding steam to drive a piston back and forth in a cylinder.

THIRD CLASS LEVER A kind of lever in which the fulcrum is at one end, the load is at the other, and the effort is in between.

TRANSMISSION The system In a vehicle that transmits motion between the engine and the driving wheels.

TURBINE An engine that uses a moving fluid (liquid or gas) to spin wheels on a shaft and produce power. Water, wind and gas turbines are now in widespread use.

TURBOFAN A jet engine that has a huge fan mounted in front. Part of the engine thrust comes from air blown around the main part of the engine.

TURBOJET The simplest kind of jet engine in which all the thrust comes from the jet exhaust from the engine.

TURBOPROP A jet engine in which turbines in the jet stream spin a propeller. Most of the thrust comes from the propeller.

TURNING A machining operation carried out on a lathe. The workpiece is cut while it turns (rotates).

VALVE Part of a machine that opens to let liquids or gases flow in or out, as in a gasoline engine.

WATERWHEEL A machine once widely used to harness the power of flowing water to drive machinery.

WEAVING The method used to make cloth by Interlacing two sets of threads at right-angles to each other.

WEDGE An example of the simple machine we call the inclined plane.

WHEEL AND AXLE One kind of simple machine. Turning a large wheel moves round a smaller one (axle), thus giving a mechanical advantage.

WINCH A power-driven mechanism for winding a rope or cable. It is a kind of wheel and axle.

WINDLASS A hand-operated winding mechanism, working on the principle of the wheel and axle.

WINDMILL A machine once widely used for driving millstones, pumps, and so on. It is a kind of turbine that spins when the wind blows through the sails.

For Further Reading

Ardley, Neil.
Science Book of Machines.
Harcourt, New York. 1992.

Bains, Rae.
Simple Machines.
Troll, Mahwah, N.J. 1985.

Erikson, Sheldon.
Machine Shop.
AIMS Educational Foundation, New York. 1993.

Gardner, Robert.
Forces and Machines.
Simon and Schuster, New York. 1991

Gifford, C.
Machines.
EDC, Tulsa. 1994.

Chant, Christopher.
Illustrated Guides Series.
Marshall Cavendish, Tarrytown, N.Y. 1990.

Lambert, Mark.
Machines.
Franklin Watts, Chicago. 1991.

Seller, Mick.
Wheels, Pulleys, and Levers.
Franklin Watts, Chicago. 1993.

Wilkin, Fred.
Machines.
Children's Press, Chicago. 1986.

Answers

Page 12
Investigation

2. When you place the second coin on the other side of the ruler, you find that the distance from the pencil to the center of each coin is the same.

3. When you have balanced the ruler again, you find that the distance from the single coin to the pencil is twice the distance from the two coins to the pencil.

These results follow from what is called the law of the lever that: the effort times the effort arm Is equal to the load times the load arm.

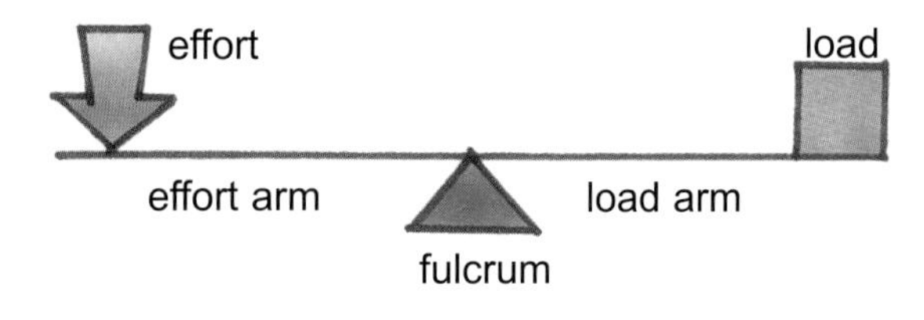

1. A pair of pliers with large jaws and small handles would not be practical. You couldn't get enough leverage with the small handles to make the big jaws grip strongly.

2. Nutcrackers are a double second-class lever; sugar tongs are a double third-class lever.

Page 14

Looking at different screws, we would say that the number of threads varies from screw to screw. But theoretically we could say that all screws have the same number of threads - just one. Every screw consists of a single spiral thread wound round a cone or cylinder.

Page 20
Investigation

The driven gear **B** turns at a speed of 11.7 rpm in the same direction as gear **A**.

Page 21

The gears must reduce the speeds transmitted by a ratio of 4 : 1.

Page 22

1. If the heat generated by friction between moving parts were not removed, the metal parts would heat up and expand. Eventually they would expand so much that they would jam together and stop moving. We say they would "seize up."

2. After braking hard, you find that the brake blocks are hot. They were heated up by friction as they pressed against the wheel rim.

3. The investigation tells you that rolling friction is much less than sliding friction. Many bearings use this principle to reduce friction between moving parts.

Page 27

Another term for crude oil is petroleum. The word "petroleum" literally means "rock oil."

Page 30

The word "hydroelectric" literally means "water-electric." It Is a good term for the electricity produced by harnessing water power.

Page 31

The space shuttle can't use jet engines to fly into space because these engines must take in air to burn their fuel.

Page 37
A space satellite doesn't slow down because there Is no air in space to push against it.

Page 41
Workout
The cotton gin could separate 300 pounds (150 kg) of cotton in a a week.

Page 43
A cutting oil is fed into the cutting area during machining for two purposes. One, it helps lubricate the cutting operation. Two, It helps remove the heat generated by friction between the cutting tool and the workpiece.

Page 45
1. If the sets of parts workers used on an assembly line were not identical, they wouldn't be able to fit them together.

2. Henry Ford first used a moving assembly line to build cars in 1913. He concentrated on building one model – the famous Model T, or "Tin Lizzie."

Page 46
If the propeller shaft was connected to the gearbox and final drive by fixed joints, they would be liable to break when the car went over bumpy roads.

Page 47
When a car goes round a corner, the outer wheels have to travel farther than the inner ones – they follow a larger radius.

Page 49
In an elevator, the weight of the passenger car is balanced by that of the counterweight on the other side of the pulley. So the elevator motor only has to provide power to lift the weight of the passengers.

Page 50
Caterpillar tracks spread the weight of the bulldozer over a larger area and give it firmer grip. They do not skid like wheels.

Page 51
Workout
On a pair of gigantic scales, it would takemore than 33 jumbo jets to balance Big Muskie.

Page 52
Fertilizers help increase food production by putting back Into the ground essential nutrients that plants take out when they grow. Pesticides help increase food production by killing pests and diseases that attack growing crops.

Page 53
Workout
It would take four workers five hours to harvest 200 acres of land.

Page 54
The science that deals with electric charges is called electrostatics.

Index